LEARNING CONTROL

LEARNING CONTROL

Applications in Robotics and Complex Dynamical Systems

Edited by

DAN ZHANG
York University
Toronto, ON, Canada

BIN WEI
Algoma University
Sault Ste Marie, ON, Canada

ELSEVIER

Elsevier
Radarweg 29, PO Box 211, 1000 AE Amsterdam, Netherlands
The Boulevard, Langford Lane, Kidlington, Oxford OX5 1GB, United Kingdom
50 Hampshire Street, 5th Floor, Cambridge, MA 02139, United States

Library of Congress Cataloging-in-Publication Data
A catalog record for this book is available from the Library of Congress

British Library Cataloguing-in-Publication Data
A catalogue record for this book is available from the British Library

ISBN: 978-0-12-822314-7

Publisher: Matthew Deans
Acquisitions Editor: Dennis McGonagle
Editorial Project Manager: John Leonard
Production Project Manager: Poulouse Joseph
Designer: Victoria Pearson

Typeset by VTeX

Contents

List of contributors

Simge Akay
Computer Engineering Department, Bahcesehir University, Istanbul, Turkey

Basim Alghabashi
Concordia Institute for Information Systems Engineering (CIISE), Concordia University, Montreal, QC, Canada

Mohamed Al Mashrgy
Electrical and Computer Engineering (ECE), Al-Mergib University, Alkhums, Libya

Nafiz Arica
Computer Engineering Department, Bahcesehir University, Istanbul, Turkey

Zeinab Arjmandiasl
Concordia Institute for Information Systems Engineering (CIISE), Concordia University, Montreal, QC, Canada

Muhammad Azam
Electrical and Computer Engineering (ECE), Concordia University, Montreal, QC, Canada

B. Balasingam
Department of Electrical and Computer Engineering, University of Windsor, Windsor, ON, Canada

Jamal Bentahar
Concordia Institute for Information Systems Engineering (CIISE), Concordia University, Montreal, QC, Canada

F. Biondi
Faculty of Human Kinetics, University of Windsor, Windsor, ON, Canada

Aaron Boda
Department of Earth and Space Science and Engineering, Lassonde School of Engineering, York University, Toronto, ON, Canada

Nizar Bouguila

Concordia Institute for Information Systems Engineering (CIISE), Concordia University, Montreal, QC, Canada

Duygu Cakir

Computer Engineering Department, Bahcesehir University, Istanbul, Turkey

Mark Green

Faculty of Science, Ontario Tech University, Oshawa, ON, Canada

Sorin Grigorescu

Robotics, Vision and Control (ROVIS), Transilvania University of Brasov, Brasov, Romania

Artificial Intelligence, Elektrobit Automotive, Brasov, Romania

Baoxin Hu

Department of Earth and Space Science and Engineering, Lassonde School of Engineering, York University, Toronto, ON, Canada

Xishi Huang

RS Opto Tech Ltd., Suzhou, Jiangsu, China

M. Khalghollah

Schulich School of Engineering, University of Calgary, Calgary, AB, Canada

Howard Li

Department of Electrical and Computer Engineering, University of New Brunswick, Fredericton, NB, Canada

C.J.B. Macnab

Schulich School of Engineering, University of Calgary, Calgary, AB, Canada

Narges Manouchehri

Concordia Institute for Information Systems Engineering (CIISE), Concordia University, Montreal, QC, Canada

Jun Meng

Zhejiang University Robotics Institute, Hangzhou, China

Afshin Rahimi

Department of Mechanical, Automotive and Materials Engineering, University of Windsor, Windsor, ON, Canada

P. Ramakrishnan
Department of Electrical and Computer Engineering, University of Windsor, Windsor, ON, Canada

Jing Ren
Department of Electrical and Computer Engineering, Ontario Tech University, Oshawa, ON, Canada

Jianguo Wang
Department of Earth and Space Science and Engineering, Lassonde School of Engineering, York University, Toronto, ON, Canada

Jin Wang
Zhejiang University Robotics Institute, Hangzhou, China

CHAPTER 1

A high-level design process for neural-network controls through a framework of human personalities

M. Khalghollah and C.J.B. Macnab
Schulich School of Engineering, University of Calgary, Calgary, AB, Canada

Contents

1.1 Introduction

This work addresses the difficulty in designing a learning control for system dynamics that contain both a large bias and significant nonlinearities, using the example of a robotic manipulator to develop and test the ideas. The vast majority of learning-control techniques in the literature addressing tracking control of the robot's end-effector try to cancel the force of gravity, or compensate for velocity-dependent nonlinearities that tend to pull the tip off-track, i.e. friction, Coriolis, and centripetal (FCC) terms. The ad hoc

Learning Control
https://doi.org/10.1016/B978-0-12-822314-7.00006-7

methods found in the literature that proposed learning both gravity and FCC terms end up doing so sequentially, not simultaneously, to the best of our knowledge. In this paper, we suggest the basic difficulty in trying to achieve both at the same time stems from the need to approach these two problems differently, i.e. they require two different types of learning.

Gravity acting on a robot qualifies as a nonlinear term with significant bias; for a simpler example of a *biased nonlinearity*, consider a cosine term near the origin. Control systems generally endeavor to cancel biased nonlinearities with a *feedforward term*, i.e. an open-loop, bias-compensation, or set-point control. A sine function, on the other hand, provides a simple example of a nonlinear function with a value of zero at the origin; compensating for this type of *unbiased nonlinearity* requires nonlinear feedback. A learning control for this problem essentially ends up achieving a *memorized feedback term*. Since the FCC terms in a robot manipulator go to zero when the velocity goes to zero, and the need for high-speed precision remains relatively rare in industrial applications, as a practical matter the trajectory-tracking control problem without gravity ends up more like learning sine than cosine. One important difference from the designer's point of view lies in the choice of robust weight update method, i.e. designing updates for the *weights* (adaptive parameters) that remain robust to disturbances in the sense of limiting *weight drift* (overlearning) and preventing *bursting* (a sudden increase in error after a period of convergence). For unbiased nonlinearities, a *leakage term* forgetting factor that tries to drive the weights toward zero works well in practice. However, such a term would directly result in a steady-state error for biased nonlinearities; for biased nonlinearities the field of adaptive control offers deadzone, parameter projection, and supervised leakage as choices for the designer, all of which require some *a priori* knowledge of the system's parameters and/or expected disturbances; as a result many do not classify such methods as true learning systems.

Here we use our own previously-proposed graphical model of human personalities to examine the problem [1]. A resulting computational model, based on feedback theory, allows a prediction of the probability distribution of Myers–Briggs behavioral technique types in the human population, for the J/P and T/F pairs [2] as well as for the I/E and N/S pairs [3]. This biologically-inspired perspective allows more intuitive design methods for the high-level thinking required in today's advanced control-system applications [4]. The proposed framework provides insight into the nature of the two types of learning problems outlined above, and it suggests how to achieve them simultaneously on a robot arm, i.e. like humans do. Building on the idea of LQR control, the approach results in an optimization method for the design of all the control parameters (feedback gains and adaptation rates) in a learning controller—the first such formal design method appearing in the literature that extends the LQR approach to nonlinear systems, to our knowledge.

Robotic-manipulator dynamics contain significant nonlinearities; proposed control methods based on linear-system theory must, at the very least, assume implementation of a gravity-compensation term [5]. Actually *learning* the force of gravity should provide many advantages, including decreasing engineering-design costs and achieving real-time payload adaptation. An iterative learning control [6] has some advantages, but does not address robustness to disturbances and/or payloads. Radial basis function networks (RBFNs) can learn to estimate the gravity term in some robot manipulators [7,8] and some biped robots [9], but the method does not extend well to multi-link robots due to the *curse of dimensionality* when trying to add more inputs into an RBFN network.

The method in [10] learns both gravity and other nonlinearities, but the very-small leakage term used in order to avoid sag-due-to-gravity appears insufficient to deal with realistic disturbances. The authors of [11] proposed an RBFN method that adapts to both gravity (biased nonlinearity) and FCC terms (unbiased nonlinearities), but requires knowledge of the inertia matrix—which implies the designer would, in fact, know the gravity term. In [12], only gravity compensation gets proposed in the first step of backstepping for a flexible-joint robot, and not FCC terms. The same authors tackle learning all nonlinearities for a Baxter robot in [13], but the tracking performance requires a set of weights identified during a learning stage—in practice it would seem the learning stage would find the large biased nonlinearity and the tracking stage would fine-tune the performance by compensating for the unbiased nonlinearities.

In previous work, our research group presented a near-optimal control [14], which developed an approach to achieve a near-optimal control signal in the presence of gravity. A cerebellar model articulation controller (CMAC) [15], with advantages of a fast adaptation rate and real-time computational ability, was found to have unique properties for tackling this problem; freezing a set of supervisory network's local weights when the bias becomes identified can compensate for the gravity bias term, and further learning (using leakage) could then fine-tune for the FCC terms. The disadvantage was the procedure was ad hoc, based on intuitive insights into the workings of the CMAC.

Developing a formal design method, on the other hand, may have some far-reaching implications for the field of control. Some have playfully accused the field of control-systems of having a *dirty secret*: designers often choose gains and/or parameters by trial-and-error, rules-of-thumb, and/or experience. Such methods can be inadequate when facing contemporary problems of interest, such as robots interacting with unknown, unstructured environments. Optimization methods promise a path to solve this problem, but standard higher-level cognitive frameworks for the design of cost-functionals remain an open problem to our knowledge. Even in the case of the linear quadratic regulator (LQR), current theory does not provide the designer with a method for choosing the Q and R weighting matrices. For more advanced nonlinear systems that interact with an environment, researchers struggle with even creating a suitable cost-functional at the moment. This work provides a biologically-inspired method for designing cost-functionals and the value of the weightings. We use a model of human personalities; choosing a personality directly results in a choice of numerical weightings. Thus, a control system designer can use their intuitive understanding of human personalities at the high-level design stage. Not only can this avoid trial and error in picking parameters, but it may also significantly reduce the total number of parameters needed compared to manual designs. Consider an analogy to how fuzzy logic proved quite a time-saving invention for control design, as a result of allowing human intuition to guide the design of computational decision-making and reducing the number of parameters that the designer needs to choose; fuzzy control ended up significantly broadening the field of computer automation, since many more problems would lend themselves to a cost-effective and/or time-efficient design solution.

This chapter first gives a background on CMAC neural network and direct adaptive control methods, as applied to both biased and unbiased nonlinearities. The Background section ends with a short introduction to our personality theory, and describes how the framework enables design of PID+bias controls using a nonlinear quadratic regulator. In the Proposed Methods section, we show how to extend the approach to designing an adaptive learning control. The Results section illustrates a formal engineering design procedure, based on constraints and objectives, for a simple mass-spring-mass and then a two-link robotic manipulator.

1.2 Background

1.2.1 The CMAC associative-memory neural network

This work assumes an associative memory will serve as the nonlinear approximator without loss of generality, where a weighted sum of basis functions gives the estimate of nonlinear function $f(\mathbf{q})$ as

$$\hat{f}(\mathbf{q}) = \boldsymbol{\Gamma}(\mathbf{q})\mathbf{w}, \tag{1.1}$$

where $\mathbf{q} \in \mathcal{R}^n$ contains the n inputs, the weights reside in column vector $\mathbf{w} \in \mathcal{R}^N$, and $\boldsymbol{\Gamma} \in \mathcal{R}^{1\times N}$ denotes a row vector of basis functions. Typically one uses Gaussians as the basis functions in a radial basis function network (RBFN). Since the RBFN's *curse of dimensionality* makes it difficult to use for more than three inputs, the learning controls in this paper use a Cerebellar Model Articulation Controller (CMAC) instead. The CMAC consists of hypercube-domain basis functions, or *cells*, constructed with rectangles, triangles, or splines. The local nature of the CMAC cells ensures only a small number of *activated* cells require calculations, along with their corresponding weights, during any one time-step of the control.

One could build a CMAC using m offset layers of n-dimensional arrays. The input activates (indexes) one cell per layer. The construction of the basis functions in a spline-CMAC follows

$$\gamma_i = \begin{cases} \prod_{k=1}^{n}(s_k^2(q_i) - 2s_k(q_i)^3 + s_k(q_i)^4) & \text{if input } \mathbf{q} \text{ activates cell } i, \\ 0 & \text{otherwise,} \end{cases} \tag{1.2}$$

where $0 \leq s_k(q_i) \leq 1$ denotes the normalized position within the cell on the kth input; normalization produces

$$\Gamma_i = \frac{\gamma_i}{\sum_{j=1}^{m}\gamma_j}. \tag{1.3}$$

An example of a spline CMAC with normalized basis functions in one dimension appears in Fig. 1.1: for two dimensions see Fig. 1.2. Rather than an impractical allocation of memory for n-dimensional arrays, the CMAC uses hash-coding to map virtual cells to a one-dimensional physical memory, since relatively few cells ever get activated in high-dimensional space during trajectory tracking of a robot [16] (hash collisions become possible, however unlikely).

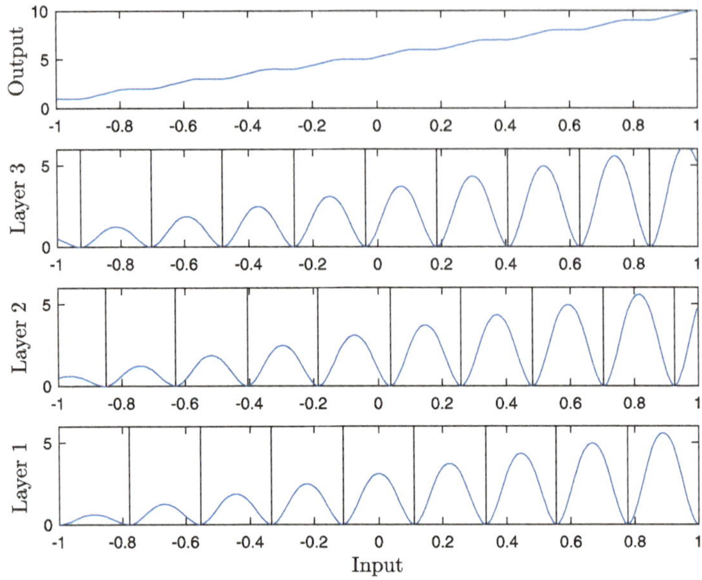

Figure 1.1 One-input CMAC with spline membership functions: $m = 3$ layers and $Q = 10$ discretizations. The output is a weighted sum of basis functions.

Being a universal nonlinear approximator, the CMAC can approximate nonlinear function $f(\mathbf{q})$ as

$$f(\mathbf{q}) = \mathbf{\Gamma}(\mathbf{q})\mathbf{w} + d(\mathbf{q}), \tag{1.4}$$

where $d(\mathbf{q})$ defines the approximation error in the uniform approximation region $D \subset \mathcal{R}$: specifically $|d(\mathbf{q})| \leq d_{\max} \forall \mathbf{q} \in D$ for a positive constant $d_{\max}$.

1.2.2 Unbiased nonlinearities

In this work the learning control strategies come from mathematical techniques developed in the field of direct adaptive control. Consider a physical system with position x and control input u

$$\ddot{x} = a_1 x + b_1 \dot{x} + f_1(x, \dot{x}) + hu, \tag{1.5}$$

where the unbiased nonlinearity has $f_1(x, 0) = 0$, with constant parameters a_1, b_1, and $h > 0$. Given desired trajectory $x_{\text{des}}(t)$, $\dot{x}_{\text{des}}(t)$, $\ddot{x}_{\text{des}}(t)$, then defining state errors $e_1 = x - x_{\text{des}}$, $e_2 = \dot{x} - \dot{x}_{\text{des}}$ and auxiliary error $z = \Lambda e_1 + e_2$ leads to an expression of the dynamics suitable for control-system design

$$\dot{z} = ax + b\dot{x} + f_2(e_1, e_2, x_{\text{des}}, \dot{x}_{\text{des}}, \ddot{x}_{\text{des}}) + hu. \tag{1.6}$$

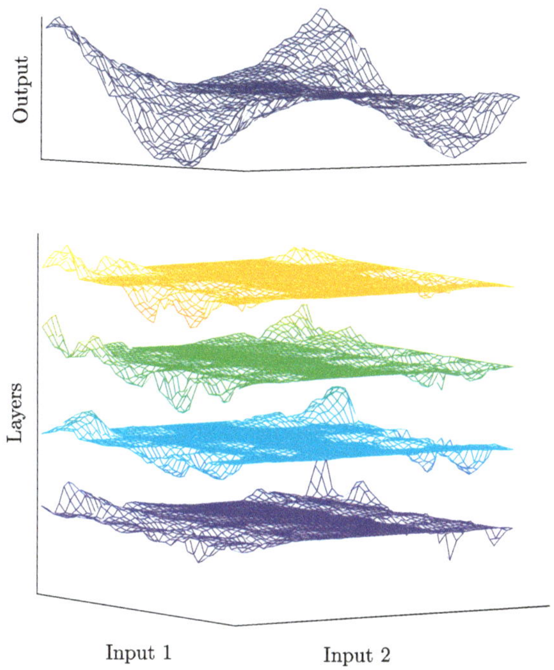

Figure 1.2 Normalized-spline, two-input CMAC: $m = 4$ and $Q = 4$. The word *layers* for a CMAC refers to hash-coded arrays indexed in parallel.

Although a desired trajectory may introduce a system bias, such that $f_2(e_2 = 0) \neq 0$, a practical neural-adaptive control ignores this typically-small effect and a common design uses leakage [17] multiplied by positive parameter ν in the robust weight update

$$u = -\mathbf{\Gamma}\hat{\mathbf{w}} - K_d z, \tag{1.7}$$

$$\dot{\hat{\mathbf{w}}} = K_i(\mathbf{\Gamma}^T z - \nu\hat{\mathbf{w}}), \tag{1.8}$$

where $\hat{\mathbf{w}}$ indicates the ideal-weight estimates, positive parameter ν determines the influence of the robust leakage modification, which limits weight drift by *trying to force* the weights to zero. As long as $|f_2|$ remains relatively small, the leakage term does not significantly reduce performance. The definition of z implies positive constant K_d equates to a derivative gain from PID control—giving an effective proportional gain $K_p = \Lambda K_d$. We denote the positive-constant adaptation gain as K_i because in subsequent sections will we point out similarities of the neural network to an integral term, i.e. this work treats a neural network trained with adaptive-control update laws as just a *memorized integral*.

Note that the system control actually occurs as a multi-rate signal, with the feedback running fast enough to approximate a continuous time signal and the neural network updated at a discrete rate, i.e. a *zero-order-hold* discretized signal describes its output characteristics. Designers often choose leakage for a robust update, and in discrete time with sampling period Δt the delta-rule update becomes

$$\Delta\hat{\mathbf{w}} = \Delta t K_i(\mathbf{\Gamma}^T z - \nu\hat{\mathbf{w}}), \tag{1.9}$$

Note the delta-rule simply results in the neural network estimating the discrete-time model rather than the continuous-time model using its universal approximation ability, i.e. it does not cause instability when Δt is relatively small [18].

1.2.3 Direct adaptive control in the presence of bias

Consider adding a biased nonlinearity, $g(x)$ with $g(0) \neq 0$, to the previous model

$$\dot{z} = ax + b\dot{x} + f_2(e_1, e_2, x_{\text{des}}, \dot{x}_{\text{des}}, \ddot{x}_{\text{des}}) + g(x) + hu. \tag{1.10}$$

In this case the leakage term multiplied by ν in (1.8) typically *pushes* the weights in the wrong direction and performance suffers; deadzone provides a more practical robust modification

$$\dot{\hat{\mathbf{w}}} = \begin{cases} 0 & \text{if } |z| < \epsilon_{\max}/K_d, \\ \beta\mathbf{\Gamma}^T z, & \text{otherwise,} \end{cases} \tag{1.11}$$

where $\epsilon_{\max}$ includes the maximum approximation error of the neural network within its domain $d_{\max}$, combined with maximum external-disturbance amplitude. Thus, the amount of system-plus-disturbance knowledge required may make the method impractical in many applications.

Many papers in the literature suggest adaptive-parameter projection instead, even simple versions like

$$\dot{\hat{\mathbf{w}}}_i = \begin{cases} 0 & \text{if } \hat{w}_i > w_{i,\max} \text{ and } \mathbf{\Gamma}^T z > 0, \\ 0 & \text{if } \hat{w}_i < w_{i,\min} \text{ and } \mathbf{\Gamma}^T z < 0, \\ \beta\mathbf{\Gamma}^T z & \text{otherwise.} \end{cases} \tag{1.12}$$

The designer requires knowledge of the maximum and minimum weight for the i cell, $w_{i,\max}$ and $w_{i,\min}$ respectively, presumably identified during pre-training.

Without *a priori* knowledge of the system and disturbances, learning control becomes the logical choice in the context of biased nonlinearities. A direct adaptive control using supervised leakage,

$$\dot{\hat{\mathbf{w}}} = K_i \mathbf{\Gamma}^T z + K_b(\bar{\mathbf{w}} - \hat{\mathbf{w}}), \tag{1.13}$$

can incorporate a learning control scheme when the weights that estimate the bias $\bar{\mathbf{w}}$ are output from an online algorithm attempting to identify the bias in real time; for stability reasons $\bar{\mathbf{w}}$ should only be updated once. The robust term contains multiplying parameter denoted $K_b = K_i \nu$ because in subsequent sections will show this term has similar effects as a control-bias term (a setpoint or feedforward term).

1.2.4 A graphical model of personalities

The graphical model of personalities consists of a quadrant with empowerment/manipulation on the horizontal axis, indicating how one tries to affect the world and people, and emotion/logic on the vertical axis indicating how one makes decisions. Manipulation includes commands, requests, charm, and all subtle forms of the art. Empowerment includes running away, avoiding, giving away power—and at higher levels of thinking includes creating safe structures, teaching, learning, planning, among many other things that do not clearly come across as manipulation. The vertical axis denotes the emotion and logic. Emotion provides coded memories, felt rather than remembered, for quick access and for giving useful direction. By logic we mean *deductive logic*, based on a set of beliefs; recall logic qualifies as either valid or invalid, whereas beliefs classify as either consistent or inconsistent. For instance, two people can disagree all while following perfectly logical arguments based on their own beliefs. Also, people only have a (perhaps delusional) self-image of performing valid logical deductions and predictions.

Eight of the clinical personality disorders (PDs) anchor the system, since they exhibit widely-agreed-upon qualities. Histrionic and Borderline lie in the upper-right (UR), Avoidant and Dependent in the upper-left (UL), Schizoid and Schizotypal sit in the lower-left (LL), and finally the lower-right (LR) has Narcissistic and Antisocial. The characteristic on the axis to the counter-clockwise for each personality disorder seems quite apparent to all as the main behavior; everyone experiences URs as emotional, ULs are seen to empower others by leaving them alone or surrendering all their power, LLs are seem overly logical when they are creating theories (even if the underlying belief system seems unique), and LRs have reputations for their shameless manipulation. The hidden characteristic becomes apparent as a main motivator typically only to people who engage in intimate relationships with PDs. Inside the circle indicated in Fig. 1.6, better-balanced people have *personality self-images* rather than PDs; for personalities the visible and hidden characteristics predict biases rather than behaviors. The visible axis indicates a *projected bias*, felt as an irrational hope; we describe the hidden as a *reflected bias*, felt as an irrational fear of others exhibiting this characteristic. Note that one does not necessarily move outwards on the graph to a PD in one's quadrant if one was having mental health issues [19], [20], and our model does not yet predict movement on the graph. Personality theories that indicate self-images include Lowry Colors and the Enneagram.

The Lowry True Colors system [21] provides personality descriptions that closely match the mentioned qualities (Fig. 1.6). Briefly, the *Green* personality types feel primarily logical; many engineering undergraduate students feel Green. The *Gold* personality types feel mission-oriented and like to empower others through sharing knowledge and meticulous planning; many teachers experience themselves as Gold. The *Orange* personality types feel action-oriented and like to get things done by leading (manipulating) others; many managers perceive themselves as Orange. The *Blue* personality types feel primarily people-oriented and focus on building relationships using their emotions; many empathy-based therapists experience themselves as blue.

The Enneagram has found vast popularity with the general public and has nine types. If we move the types 1, 4, and 8 into new positions it achieves consistency with the quadrant system. Full qualitative descriptions of how the personality disorders, Lowry Colors, and Enneagram personalities fit on the self-image quadrant appear in [1]. Also by mapping Merrill's C.A.P.S. theory of work place personalities [22] onto the quadrant, when considering it as the way others perceive someone. For example, opposite types in the Enneagram come across similarly to the outside world because opposite self-images also imply opposite unconscious and opposite unconscious communications.

The quadrant matches closely to how people in Western society anthropomorphize some animals; a biological explanation for personalities appar-

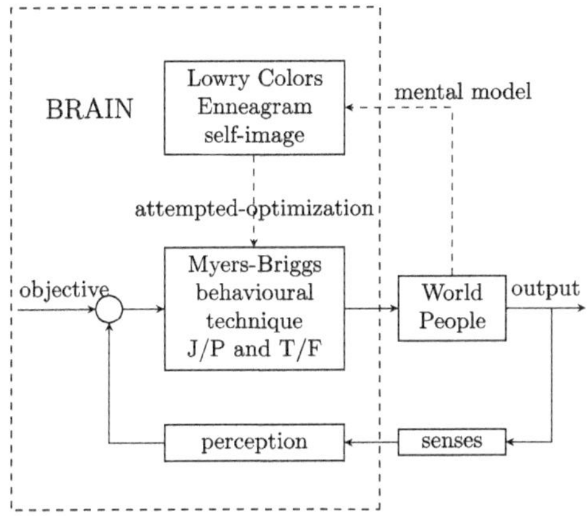

Figure 1.3 Feedback model used to explain the probability distribution of J/P and T/F in the human population in [2].

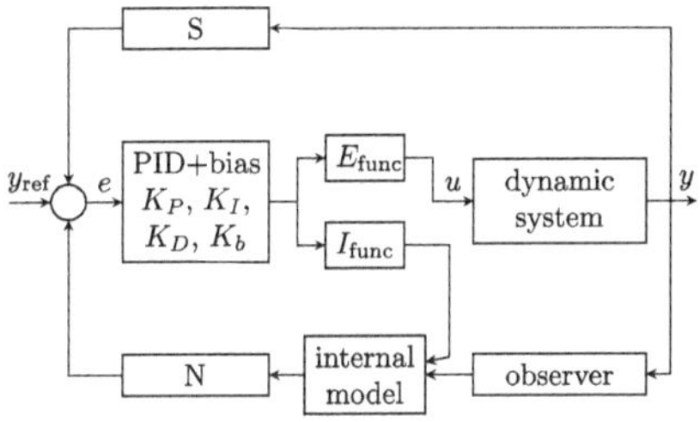

Figure 1.4 Control-system model from [3] used to predict I/E and N/S distributions.

ently becomes a possibility if we imagine that, earlier in evolution, empowerment may have started with just *passive* behavior. Manipulation may have started with simple *active* (Fig. 1.5) movements. What if we consider animals as a type of robot, engaging with the world through feedback? Then active may indicate closed-loop control, and passive open-loop control. For a human example, think of an athlete learning a new movement; at first it is an active movement that requires closed-loop control relying on feedback, with resulting excessive effort. Constant practice and training will eventually turn the movement into a memorized passive movement using open-loop control, resulting in little (minimal) effort due to relaxed muscles.

1.2.5 A computer model of personalities

When the sensors provide measurements of the system error e and its derivative $\dot{e}$, most industrial applications use a traditional PID+bias con-

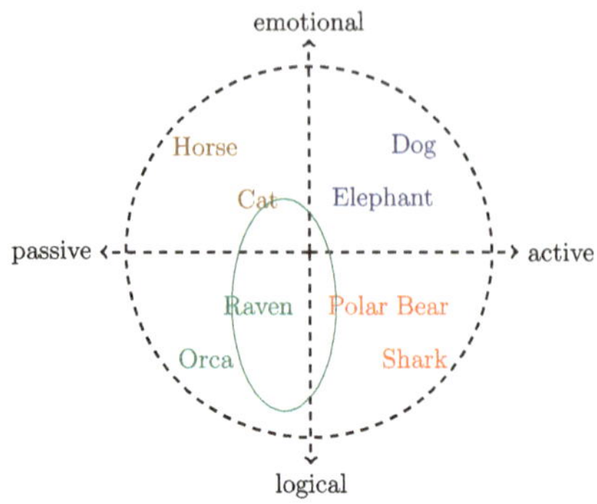

Figure 1.5 Personality self-images of Lowry True Colors can be understood by anthropomorphizing animals. Green ellipse: many engineering students may identify with Orcas who learn logical hunting techniques, which may be called passive since they are very careful about what they eat.

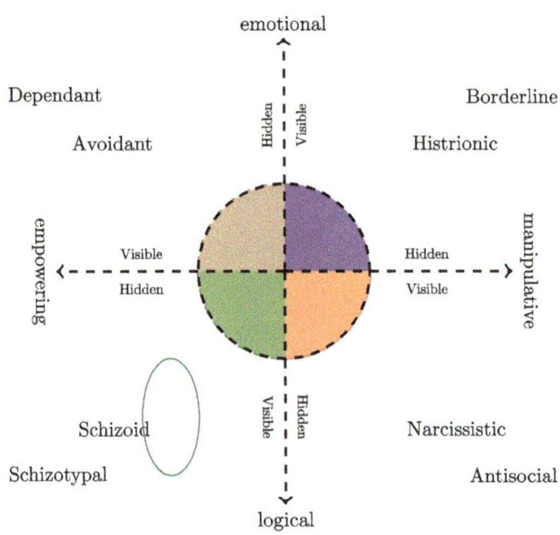

Figure 1.6 The clinical personality disorders represent unbalanced people. Green ellipse: a shizoid tells people about their theories (visible logic), with the idea of warning, helping, or enlightening others (hidden empowerment).

trol

$$u = -K_P e - K_D \dot{e} - K_I \int e \, dt + K_B, \tag{1.14}$$

where the constants K_P, K_D, and K_I defines the control gains, and K_B defines the control bias. In order to design optimal gain parameters, one must choose a cost-functional. The reader may already know how to use an LQR cost-functional to pick optimal PD gains; here we use a nonlinear quadratic regulator cost functional based on a framework of personality self-images

$$\begin{aligned} \text{Cost} = Q_1 \times (\text{Total Position}) + Q_2 \times (\text{Total Velocity}) \\ + R_1 \times (\text{Total Effort}) + R_2 \times (\text{Total Learning Error}), \quad (1.15) \end{aligned}$$

where the words indicate overall norm measurements e.g. $\mathcal{L}_2$ norms; the first two terms equate to the terms multiplied by the **Q** matrix in an LQR control (assuming an SISO system with position and velocity states), while the third R_1 term penalizes the PID control effort for the first half of the step response, but for the second half acts like traditional LQR only penalizing a measure of PD feedback effort, e.g. $\int (K_P e + K_D \dot{e})^2$. The R_2 term stems from understanding the integral term as trying to compensate for the bias (or inaccuracies in bias compensation). Thus, an integral qualifies as a simple learning term, and its *error* measures the distance from its ideal value (the bias K_B that would result in $u \equiv$ bias as the error reaches zero at steady state). We point out the following similarities to the qualities in the graphical theory of personalities:

$$\text{empowerment} \longleftrightarrow R_1, \qquad \text{manipulation} \longleftrightarrow Q_1,$$
$$\text{emotion} \longleftrightarrow R_2, \qquad \text{logic} \longleftrightarrow Q_2,$$

which allows a high-level control design e.g. choosing a desired personality results in the weighting parameters in a nonlinear quadratic regulator (NQR). This eliminates the need to pursue trial-and-error, rule-of-thumb, or experiential choices of Q's and R's, allowing an intuitive understanding of human personalities to guide the design process at the highest level (Fig. 1.7).

In previous work we pointed out similarities in PID+bias control to the Myers–Briggs personality behavioral techniques for humans:

$$\text{Perceiving} \longleftrightarrow K_B, \quad \text{Judging} \longleftrightarrow K_P,$$
$$\text{Feeling} \longleftrightarrow K_I, \quad \text{Thinking} \longleftrightarrow K_D,$$

and this model (Fig. 1.3) allowed us to predict the probability distribution of Myers–Briggs conflict types (J/P and T/F) in the human population. The optimization problem only requires three parameters, so the four-parameter PID+bias is overparameterized; in terms of human personalities we view

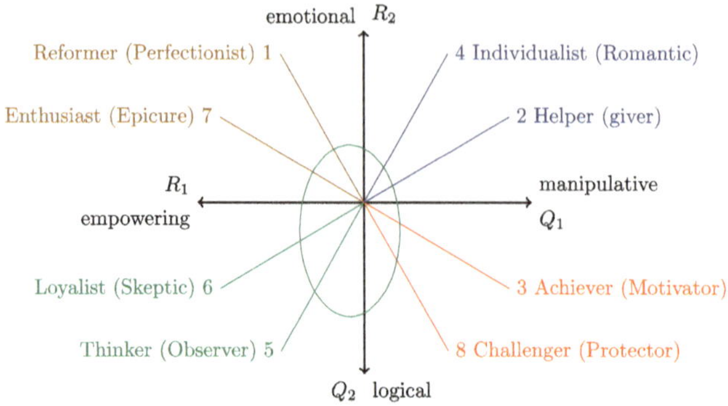

Figure 1.7 The proposed new Enneagram configuration moves types 1, 4, and 8 to make it consistent with the quadrant system. Noticing similarities between the four personality qualities and Q_1, Q_2, R_1, and R_2 in an NQR control results in an intuitive control-system design process. The oval gives way to visualize the self-image of a Type 5, for example, who has all four qualities but not in equal measure.

this as free will for an individual. Thus, the optimization must start with one *a priori* parameter which we will refer to as our *free-will parameter.*

The designer can first choose a personality, a precise angle on the personality quadrant, and then choose a personality imbalance for a magnitude. Consider using a parameter p to define the ratio of personality characteristics on opposite sides of the graph: for example in Fig. 1.8 the designer has chosen a Type 5 personality and by choosing $Q_2 = \sin(60)$ and $R_1 = \cos(60)$ the other qualities stem from the chosen imbalance, $Q_1 = pR_1$ and $R_2 = pQ_2$ with $p = 0.5$.

Consider the proposed precise design for the cost-functional

$$F = \sum_{i=1}^{2}\Bigg(\int_{\tau=0}^{T} Q_1 e^2 + Q_2 \dot{e}^2 + R_1 u_F^2 d\tau + R_1 \int_{\tau=0}^{T/2} (u_I + E_{o,i})^2 d\tau + R_2 \int_{\tau=0}^{T} L_i^2 d\tau\Bigg), \tag{1.16}$$

where $R_1 = R$ and $L_i = u^* - K_B - (u_I + E_{o,i})$ defines the learning error of the integral term. The integral effort gets penalized during the first half of the step response, but after that we worry only about its learning error. PD control defines the pure feedback component $u_F = K_P e + K_D \dot{e}$ while the memory (integral) term $u_I = K_I \int_{t=0}^{\tau} e\, dt$ uses initial condition $u_I(0) = E_o$. The algorithm sets an initial condition for the integral term as $E_{o,1} = 0$ in the first run and then at $E_{o,2} = u^*$ for the second. Note the first term

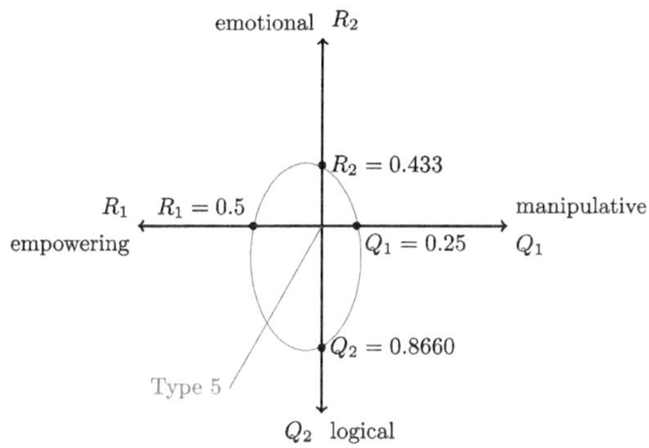

Figure 1.8 Example: choice of Type 5 personality with imbalance parameter $p = 0.5$ results in particular values for the NQR cost-functional weightings.

simply uses traditional LQR design; the NQR simply adds a component that encourages the integral's *learning error* to draw near zero by time T_f.

Thus, a nonlinear quadratic regulator producing an optimal PID+bias control models how humans choose a behavioral technique (Myers–Briggs) to try and achieve consistency with their self-image personality (Enneagram); since the model stems from the principles of feedback, the assumption that evolution did not find a way to escape the principles of feedback implies this model may indeed capture the underlying principles of human personalities. Modeling the other Myers–Briggs letters with feedback (Fig. 1.4) allowed us to predict the probability distribution of N/S (given I/E) in the human population, and the interested reader might refer to [23] for more detailed explanations.

1.3 Proposed methods

1.3.1 Proposed learning law

Based on our previous observations we propose using supervised leakage

$$\dot{\hat{\mathbf{w}}} = K_i \boldsymbol{\Gamma} z + K_b(\bar{\mathbf{w}} - \hat{\mathbf{w}}), \tag{1.17}$$

where the bias weights start at zero, $\bar{\mathbf{w}}(0) = 0$, and

$$\bar{\mathbf{w}}(t > T_c) = \hat{\mathbf{w}}(T_c), \tag{1.18}$$

where T_c the moment in time the algorithm *captures* a bias estimate, i.e. when the system first begins to remain close to the origin with close to

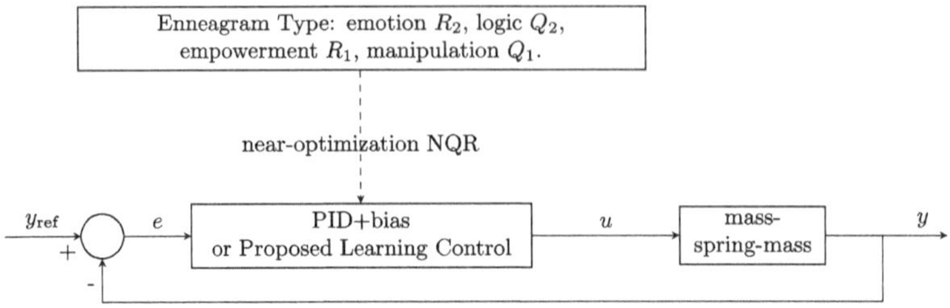

Figure 1.9 Computational model stemming from Fig. 1.3. NQR models Enneagram and PID+bias models MB conflict types (J/P and T/F).

zero velocity or, equivalently, a settling time. We only expect to capture an estimate of the bias, and the continued updates make up for inaccuracies in this bias estimation and compensate for more detailed characteristics of the nonlinearities near the origin.

1.3.2 Cost functional for optimization

Using the similarity between the four quadrant axes and the standard LQR problem, we propose the cost functional

$$\begin{aligned}\text{Cost} = & \; Q_1 \times (\text{Position Error}) + Q_2 \times (\text{Velocity Error}) \\ & + R_1 \times (\text{Feedback Effort}) + R_1 \times (\text{Extra Effort}) \\ & + R_2 \times (\text{Learning Error}). \end{aligned} \tag{1.19}$$

The first three terms look familiar, known from the LQR cost functional, while the feedback effort only includes

$$u_F = -K_p e_1 - K_d e_2, \tag{1.20}$$

and not the neural network outputs (Fig. 1.9).

The total extra effort should penalize the effort needed above a positive bias, but not any effort below the bias. Here we look at this term as an opportunity to also limit weight drift; rather than measure neural network output directly we look at the level of individual weights, and penalize each weight for being too large-positive. Otherwise, a weight activated only in a region where the basis function had a small value wouldn't get penalized enough for its growth; the proposed method ensures each weight gets equally penalized for growing regardless of the value of the basis function. This measurement of extra effort W_k gets updated at time step k, of length

Δt, according to

$$\Delta W_k = \begin{cases} \hat{w}_j - \bar{w}_j & \text{if } \hat{w}_j > \bar{w}_j \text{ and } z > 0, \\ 0 & \text{otherwise,} \end{cases} \quad \text{for } j = 1, \ldots, m.$$

The *learning error* denotes the difference between neural network portion of the control and all the nonlinear terms; describing by functions of time gives the expression

$$E_k = -\boldsymbol{\Gamma}(k\,\Delta t)\hat{\mathbf{w}}(k\,\Delta t) - [f_2(k\,\Delta t) + g(k\,\Delta t)].$$

Unlike an LQR control, the design with an integral must consider initial conditions. Thus, we run the step response twice; the first has all weights initialized to zero and the second keeps the same weight values from the end of the first run. Thus, the total cost functional becomes

$$\text{Cost} = \sum_{i=1}^{2} \int_{\tau=t_i}^{t_i+T} (Q_1 e^2 + Q_2 \dot{e}^2 + R_1 u_F^2) d\tau + \Delta t \sum_{k=1}^{2T/\Delta t} (\eta R_1 W_k^2 + R_2 E_k^2), \tag{1.21}$$

with initial condition for the two step responses $e(t_0) = e(t_1) = -1$, where $t_0 = 0$ and $t_1 = T + \Delta t$; the simulation in the Results section uses $\eta = 5$.

1.3.3 Stability analysis

This section investigates stability theory in terms of Lyapunov functions. The optimization values Q_1, Q_2, R_1 and R_2 will dive form the personality angle and magnitude; themselves functions of the personality imbalance parameters (see Fig. 1.8 for an example). The imbalance in the personality gets described by parameter p. Using the Matlab® optimization tool fmincon() generates the optimal values for K_d, K_i and K_b with the proportional gain as input, i.e. a *free will parameter*.

A Lyapunov candidate Eq. (1.22) enables stability boundary identification when the designed control law (1.7) and the proposed update term (1.13) gets applied to (1.5). The adaptive Lyapunov functions

$$V = \frac{1}{2} z^2 + \frac{1}{\beta} \tilde{\mathbf{w}}^T \tilde{\mathbf{w}}, \tag{1.22}$$

where $\tilde{\mathbf{w}} = \mathbf{w} - \hat{\mathbf{w}}$ indicates the weight errors, with column vector $\mathbf{w}$ indicating the (unknown) ideal values of the weights, results in time derivative

$$\dot{V} = -K_d z^2 + z(\epsilon + d) - \frac{K_b}{K_i}\tilde{\mathbf{w}}^T(\bar{\mathbf{w}} - \hat{\mathbf{w}}). \tag{1.23}$$

Those familiar with (supervised) leakage might recall that (1.23) implies signals are uniformly ultimately bounded (see the appendix).

1.4 Results

1.4.1 Developing a design procedure

Consider the vertical mass–spring–mass system

$$m_1\ddot{y} + b_1\dot{y} + k(y - x) + f(y) = u - m_1 g, \tag{1.24}$$
$$m_2\ddot{x} + b_2\dot{x} + k(x - y) = -m_2 g, \tag{1.25}$$

where y and x describe the positions of the two masses $m_1 = 1/g$ and $m_2 = 0.1$, $b_1 = 0.1$, $b_2 = 0.05$ denote friction coefficients, $k = 100$ gives the spring constant, $g = 9.81$ describes gravitational acceleration, and u provides the input force. Note $u^* = (m_1 + m_2)g$ so that the error $e = y$ ensures regulation about the position $y = 0$, i.e. gravity forms part of the biased nonlinearity. The second equation represents unmodeled dynamics. The nonlinearity, including both biased and unbiased terms, consists of

$$f(y) = \cos(5y) + \sin(5y). \tag{1.26}$$

This method proposed a formal engineering design process for learning controls based on meeting objectives and constraints—perhaps the first of its kind ever proposed. We use the following constraints in the design, monitored during a step response:

Constraint 1) Position constraint $|e(t = 30)| < 0.05$ m,
Constraint 2) Velocity constraint $|\dot{y}(t)| < 0.75$ m/s,
Constraint 3) Effort constraint $|u(t)| < 2\times$ bias.

Thus, the position constraint defines what the user would see as the effective *steady-state error* for the step response, the remaining sag below the set-point at the end of the measured step. The velocity constraint might help ensure safe operation and/or help reduce wear-and-tear of mechanical parts e.g. so that the system lasts the specified life-span as per the engineering specs. One might design the effort constraints with actuator saturation, actuator life-span, and/or safety considerations in mind. For objectives we look at minimizing total error, velocity, and/or effort over the entire step.

Objective 1) Performance objective is minimizing $\int_{t=0}^{2T} e^2$,
Objective 2) Velocity objective is minimizing $\int_{t=0}^{2T} \dot{y}^2$,
Objective 3) Effort objective is minimizing $\int_{t=0}^{2T}[(K_p e + K_d \dot{y})^2 + \mathrm{sat}(u - \mathrm{bias})^2]$.

The design of the saturation function assumes a *gravity down* scenario, so that extra effort only occurs when pushing more than the force of gravity

$$\mathrm{sat(E)} = \begin{cases} E & E > 0, \\ 0 & \text{otherwise}, \end{cases} \tag{1.27}$$

so that the effort objective measures both feedback effort and *extra* effort (beyond the bias).

We propose the following engineering design procedure:

Step 1) Start with very unbalanced personalities—choose parameter p small.
Step 2) Run step responses for all eight personality types.
Step 3) Repeat first two steps until at least three personalities meet all the constraints.
Step 4) Compare how well the constraint-satisfying controls meet the objectives.
Step 5) Choose one of these controls.

We started with the imbalance parameter $p = 0.1$. The imbalance parameter determined as a multiplier for determining the amount of personality offset. All simulations use the proportional gain as the *free will parameter* and set $K_p = 1$ for every optimization. Each optimization starts from five different random initial conditions in an attempt to find a minimum close to (or exactly on) the global minimum, for the remaining parameters K_i, K_d, K_b.

The algorithm to identify the bias used in the supervisory leakage term (1.17) was: if both the magnitude of the position error remained below 0.1 m and velocity magnitude remained below 0.1 m for 10 discrete time steps the supervisory weights $\bar{\mathbf{w}}$ were set to the current value of the weight estimates $\hat{\mathbf{w}}$.

A first attempt using $p = 0.1$ reveals no personalities meet the constraints (Fig. 1.10). But a more balanced personality with $p = 0.5$ (Fig. 1.11) has four personalities that meet the constraints; the objectives (Fig. 1.12) reveal

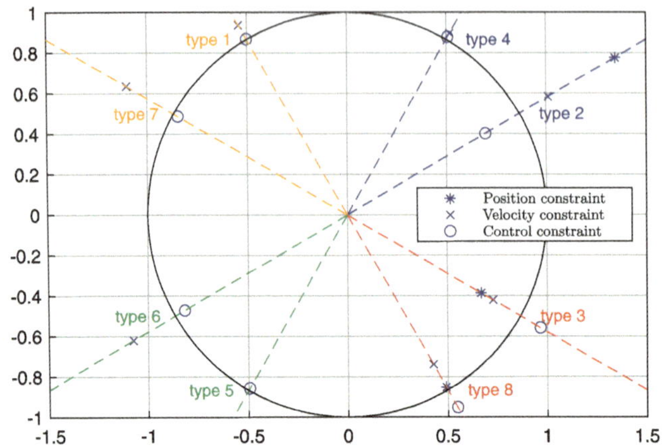

Figure 1.10 Normalized constraints with $p = 0.1$: inside the circle constraints are met. No personality met all three constraints.

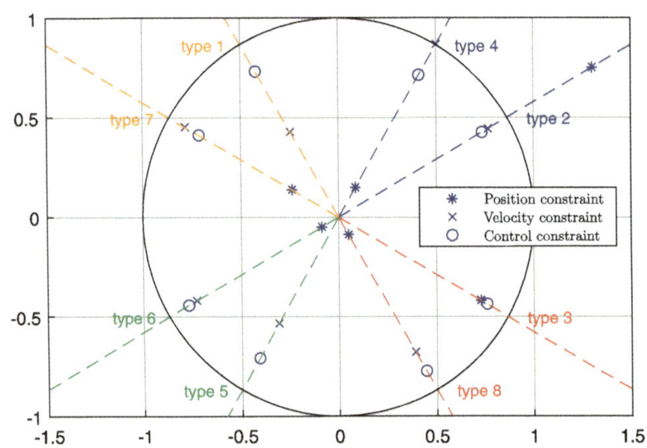

Figure 1.11 Increasing balance by using $p = 0.5$. Type 7, 6, 8 and also 3 personalities now meet all three constraints.

Type 8 does best for velocity (Fig. 1.13), Type 6 shows the best result for position performance by a very slight amount (Fig. 1.14), Type 7 shows the best result for effort and does not appear to sacrifice performance to do so (Fig. 1.15), while Type 3 does not do particularly well at anything (Fig. 1.16). Thus, it seems likely a designer would choose Type 7 based on these results, which has optimization result in $K_i = 6.8$, $K_d = 2.2$, and $K_b = 1.3$. Note these figures also compare the proposed adaptive-NQR performance to a standard PID+bias design; the PD gains came from an

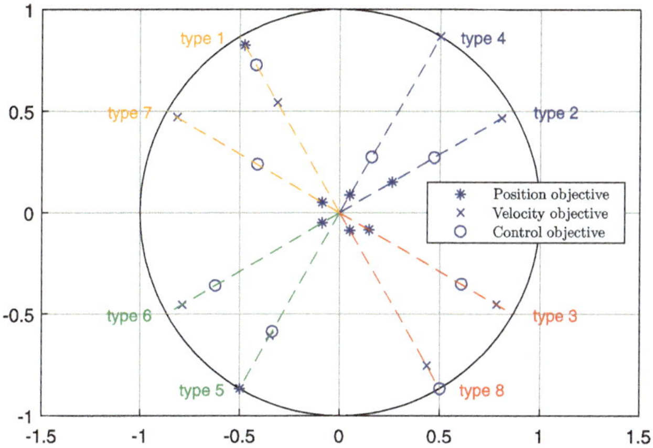

Figure 1.12 Normalized objectives with $p = 0.5$: closer to the middle is better. Of those that meet constraints, all four have similar position performance but Type 7 does best minimizing control effort.

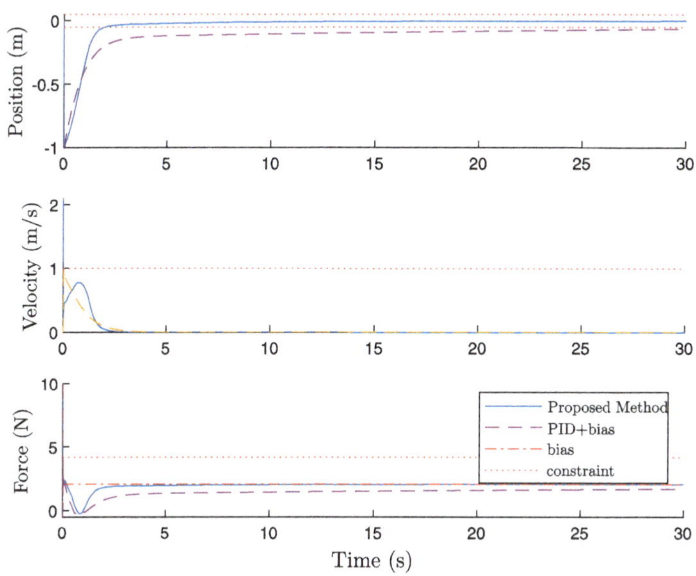

Figure 1.13 Type 8: does well on both position constraint and position objective.

LQR optimization, with the bias 0.9 of the true bias modeling (which could represent an unknown payload), and the integral gain K_I came from trial-and-error, i.e. like one might do in industrial application—the result was $K_P = 36$, $K_I = 1$, $K_D = 31$, and $K_B = 2$.

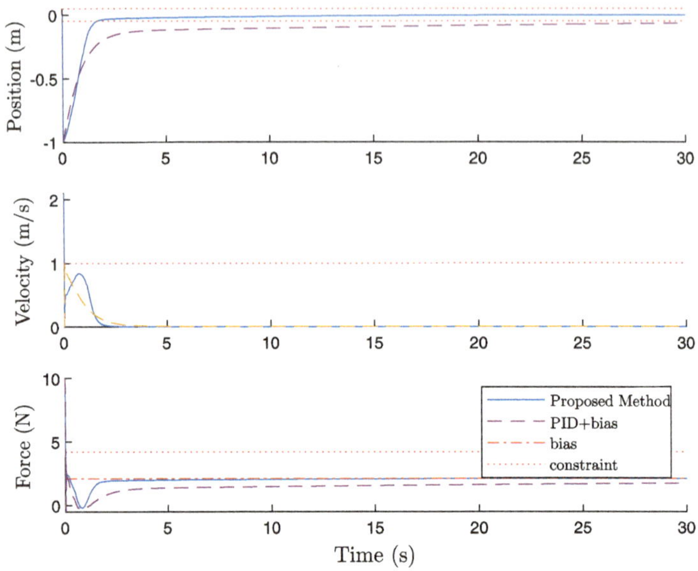

Figure 1.14 Type 6: similar performance to Type 8 but does slightly better reducing control effort.

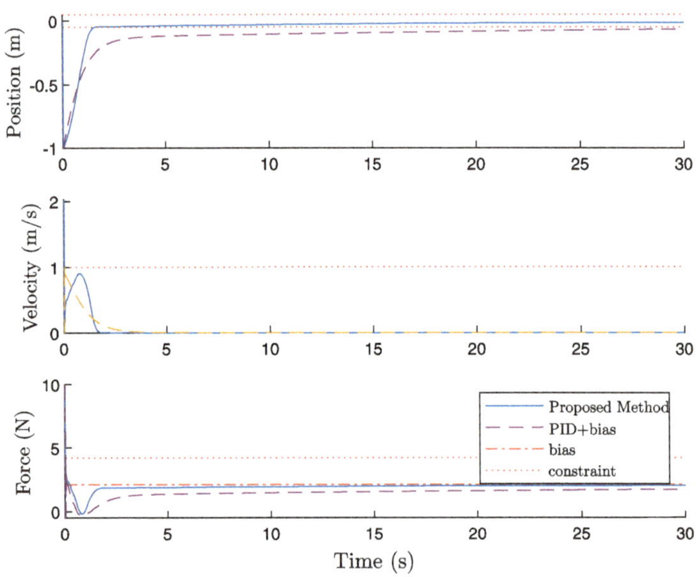

Figure 1.15 Type 7: of those that meet constraints it does the best in the minimal control-effort objective.

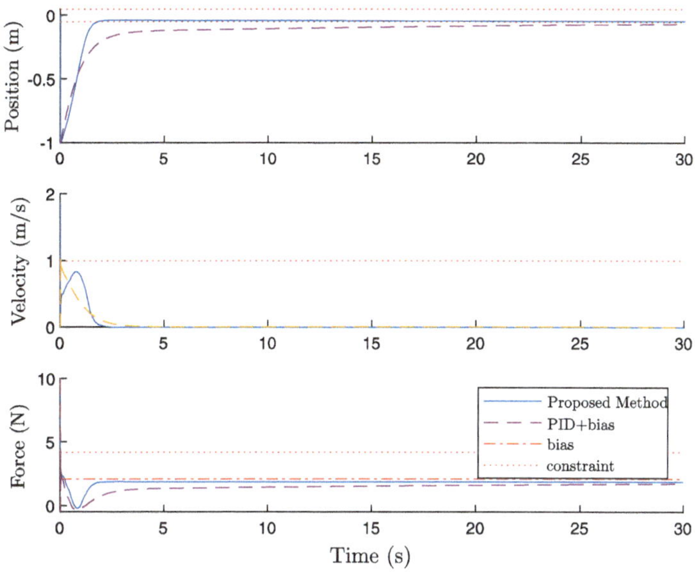

Figure 1.16 Type 3: it meets constraints but does not do particularly well in objectives compared to others.

Typically people have focused learning or adaptive methods to address either biased or unbiased nonlinearities e.g. either gravity compensation or trajectory-tracking in a robotic manipulator. Thus, here we compare our method of adapting to both kinds to addressing each individually:

Adaptive 1) $\dot{\hat{\mathbf{w}}} = \beta(\mathbf{\Gamma} z - \nu \hat{\mathbf{w}})$,
Adaptive 2) $\dot{\hat{\mathbf{w}}} = \nu(\bar{\mathbf{w}} - \hat{\mathbf{w}})$ for $t > T$,

where Adaptive (1) method uses standard leakage suitable for non-biased nonlinearities, and the Adaptive (2) method compensates for bias by using learning to find a nominal set of weights $\bar{\mathbf{w}}$ (in our simulations we update with $\dot{\hat{\mathbf{w}}} = \beta \mathbf{\Gamma} z$ until such a suitable value becomes identified at time T). Both Adaptive (1) and Adaptive (2) result in steady-state error (sag due to the bias) compared to the proposed method (Figs. 1.17 and 1.18).

1.4.2 Two-link robotic manipulator

The other simulations model involves a two-link robot arm (Fig. 1.19). Consider a rigid two-link robot arm where the X and Y describe the Cartesian coordinates of the end effector. In Table 1.1 m_1 and m_2 denote the mass related to linkage 1 and 2 respectively; l_1 and l_2 measure the distance between the 2 axes of rotation and the distance from the second motor

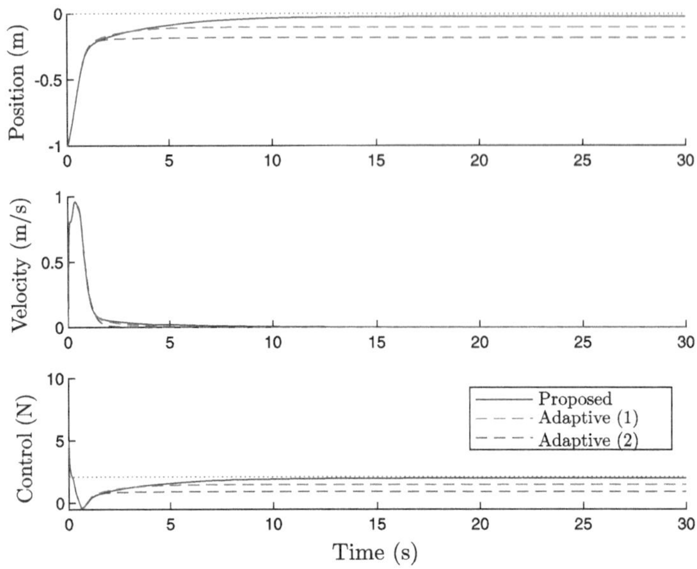

Figure 1.17 Comparing the proposed hybrid-learning method with individual learning methods: first step response in the optimization with $\hat{\mathbf{w}}(0) = 0$.

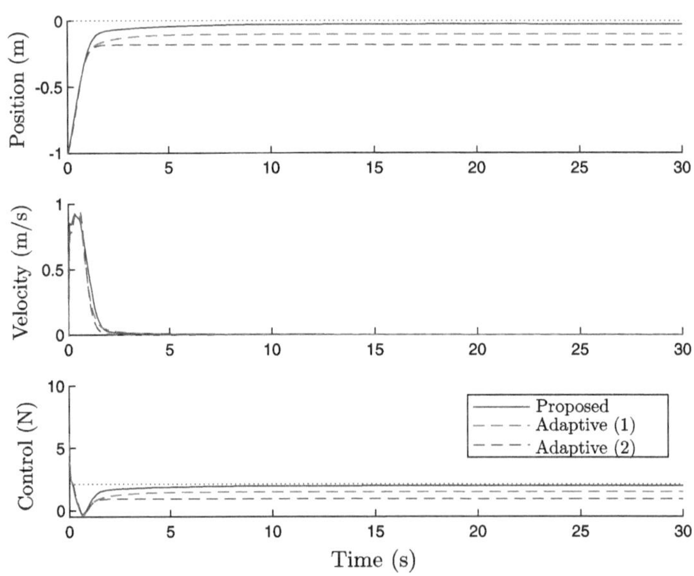

Figure 1.18 Comparing the proposed hybrid-learning method with individual learning methods: second step response in the optimization starting with the resulting weights from the first response.

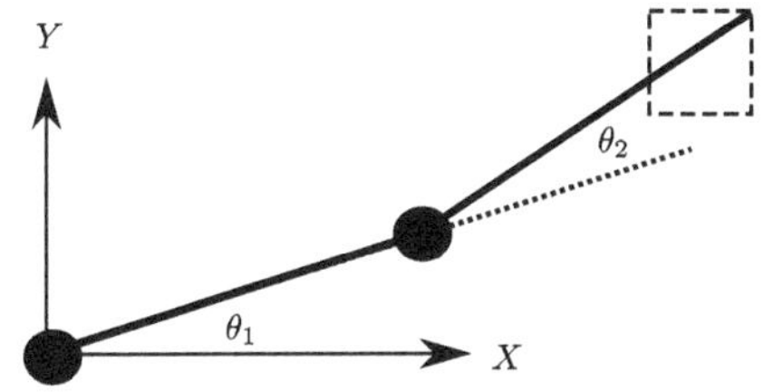

Figure 1.19 Two-link robotic manipulator, tracing a square trajectory (not to scale).

Table 1.1 The physical properties of the robot.

Parameter	Value
m_1	1.51 kg
m_2	0.87326 kg
l_1	0.34293 m
l_2	0.26353 m
J_1	0.0111 kg m^2
J_2	0.0094 kg m^2
B_1	4.5 N m s/rad
B_2	0.028211 N m s/rad

axis of rotation to the tip of link; J_1 and J_2 give the equivalent inertias of the links; and B_1 and B_2 are equivalent viscous damping coefficients of the related joints. An Euler–Lagrange formulation results in the equations of motion [24]

$$\mathbf{M}(\boldsymbol{\theta})\ddot{\boldsymbol{\theta}} + \mathbf{C}(\boldsymbol{\theta}, \dot{\boldsymbol{\theta}})\dot{\boldsymbol{\theta}} + \mathbf{D}(\dot{\boldsymbol{\theta}}) + \mathbf{G}(\boldsymbol{\theta}) = \mathbf{u} \tag{1.28}$$

where $\boldsymbol{\theta} = [\theta_1\ \theta_2]^T$ contains the joint angles, $\mathbf{M} \in R^{2\times 2}$ denotes the inertia matrix, $\mathbf{C} \in R^{2\times 2}$ contains centripetal and Coriolis terms, $\mathbf{D} \in R^2$ denotes a linear friction model, and $\mathbf{G} \in R^2$ denotes the force of gravity acting in the negative Y direction. The stability proof for controls at each joint of form (1.7) and weight updates at each joint of form (1.13) uses a Lyapunov-like lemma with Lyapunov candidate $V = \mathbf{z}^T\mathbf{M}\mathbf{z} + \sum_{i=1}^{2} \tilde{\mathbf{w}}_i^T \tilde{\mathbf{w}}_i$ where $\mathbf{z} = \boldsymbol{\Lambda}(\boldsymbol{\theta} - \boldsymbol{\theta}_{\text{des}}) + (\dot{\boldsymbol{\theta}} - \dot{\boldsymbol{\theta}}_{\text{des}})$ as in [25]), but looks very similar to the one in the appendix so we omit it here for brevity.

The design for the robot takes place with a Cartesian-space step response. We used the following design constraints: 0.01 m error for the Cartesian position, with joint velocity constraints of 2.2 and 0.05 deg/s for joint 1 and 2, respectively, and joint torque constraints of 6 and 0.5 Nm, respectively. Using an imbalance of $p = 0.1$, Types 5, 6, and 8 met all three

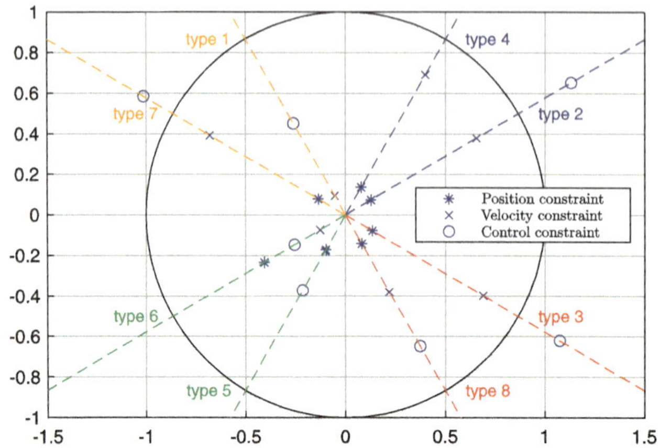

Figure 1.20 Two-joint robot arm's step response: normalized constraints using $p = 0.1$. Type 5, 6 and 8 meet all three constraints.

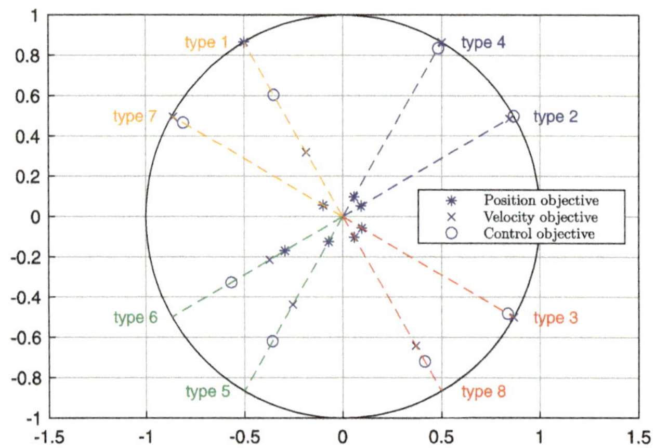

Figure 1.21 Two-joint robot arm's step response: normalized objectives using $p = 0.1$. Of the three that meet constraints, Type 8 does the best minimizing position error, while Type 6 does best minimizing velocity and control effort. Type 5 does not do particularly well.

constraints (Fig. 1.20). Type 8 did the best in minimal-overall-error objective, Type 6 did the best minimizing both overall velocity and effort, while Type 5 did not show any particularly desirable qualities in terms of meeting objectives (Fig. 1.21). Looking directly at the step responses, we can see Type 6 has quicker settling time than Type 8 (Figs. 1.22 and 1.23).

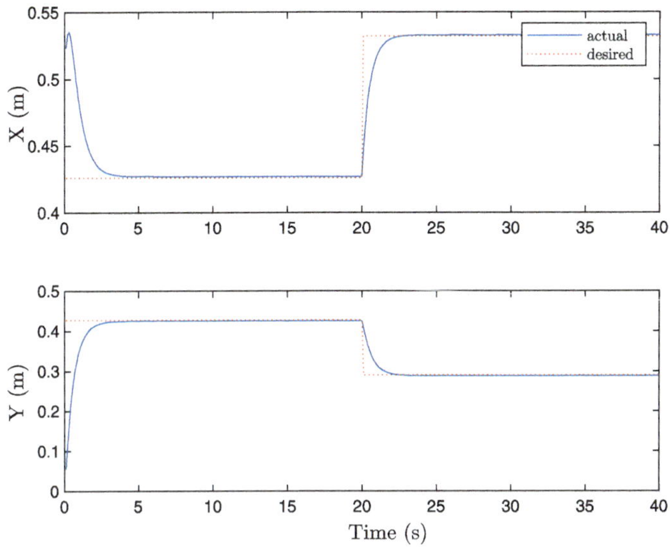

Figure 1.22 Type 6, for the Cartesian-step inputs used in the optimization-based design.

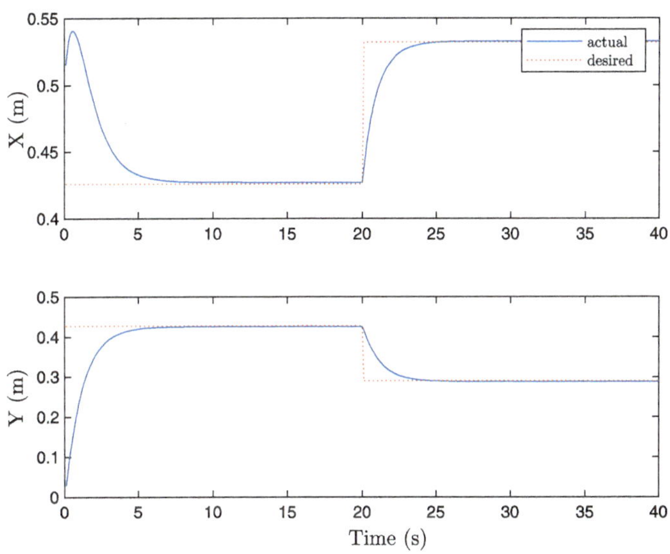

Figure 1.23 Type 8, for the Cartesian-step inputs used in the optimization-based design.

Starting with free-will parameters $K_p = 200$ for both joints, optimization for Type 6 ended up with (for joint 1, joint 1) $K_i = 350, 215$, $K_d = 1.1, 1$, and $K_b = 0.2, 0.1$ while Type 8 had $K_i = 111, 96$, $K_d = 0.9, 1$, and $K_b = 1, 1.1$. The traditional PID+bias control design (for comparison purposes) resulted in $K_P = 80, 80$, $K_I = 2, 2$, $K_D = 1, 1$, and $K_B = 1, 1$.

Another simulations test the design through tracing a square with the end effector; both Types 6 and 8 achieve very accurate control, whereas an industrial PID+bias control shows significant tracking error visible to the naked eye (Fig. 1.24 and Fig. 1.25). Figs. 1.26 and 1.27 show Type 8 uses slightly less control effort; the bottom graphs, showing RMS values for the weights, demonstrate Type 8 has the advantage in terms of output smoothness of the CMAC, which will translate into slightly smoother control signals applied to the robot. The aforementioned performances of Types 6 and 8 illustrate how well they do on the 100th repetitive trial of the trajectory; the learning curve and CMAC-weight convergence appears in Figs. 1.26 and 1.27 (Figs. 1.28, 1.29).

1.5 Conclusions

We described our previously-proposed theory of human personalities based on feedback theory, and here we detail how a designer can utilize this as a framework for high-level design process thinking in learning control systems. Choosing a particular personality, along with its imbalance, gives a phase and magnitude on our personality quadrant; this translates into values for weightings in a cost-functional for a nonlinear quadratic control (similar to LQR, but with additional terms to handle an integral term). Optimization using this cost-functional resulted in providing the gains in a PID+bias control in our previous work, while here it results in analogous parameters in a learning control system: proportional and derivative gains (as before) as well as adaptation gain and supervised leakage gain (replacing integral gain and bias). The resulting adaptation gain successfully focuses on compensation for unbiased nonlinearities, while the supervised leakage gain focuses on assisting the learning of the biased nonlinearity.

In our simulations we show how one can start with given constraints and objectives for a two-link robot; starting with a large imbalance and gradually making it more balanced can give a few personalities that meet constraints. One can then compare these few personalities for how well they meet objectives in order to select one solution. Both constraints and objectives get represented on the personality quadrant, so the whole pro-

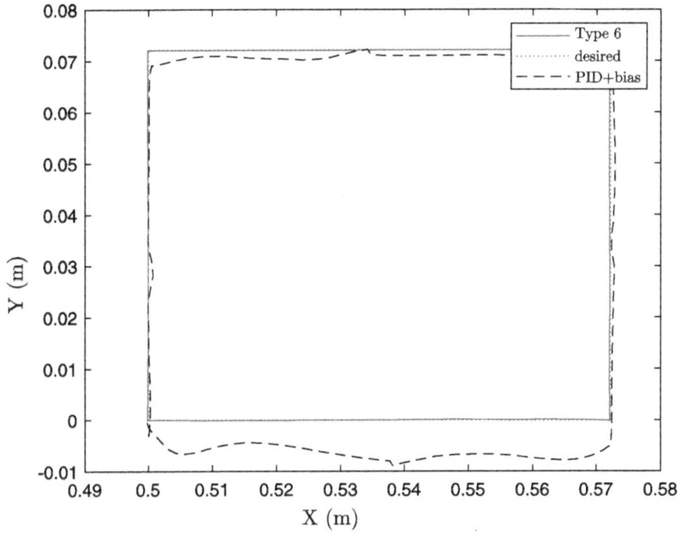

Figure 1.24 End-effector trajectory tracking after 100 repetitive trials for Type 6.

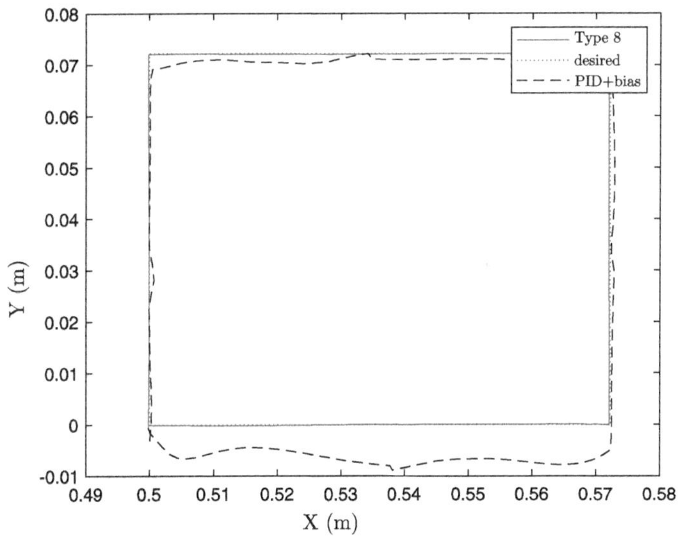

Figure 1.25 End-effector trajectory tracking after 100 repetitive trials for Type 8.

cedure becomes a graphical design method of the kind that may appeal to control-system engineers. The resulting learning system can learn both biased gravity and (nearly) unbiased FCC nonlinearities simultaneously in

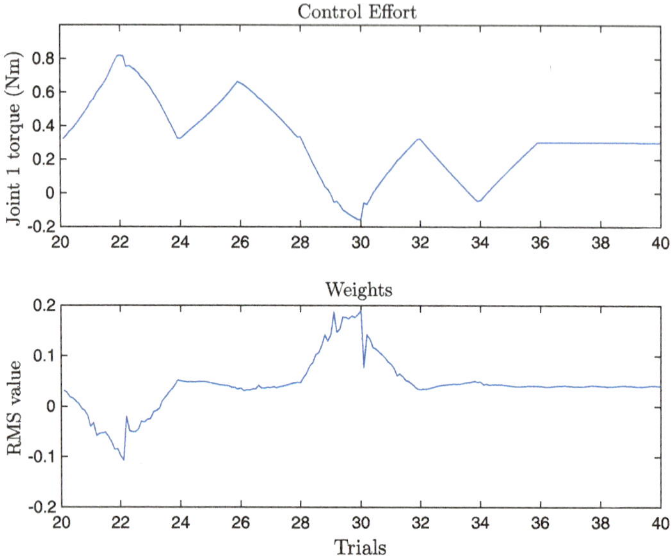

Figure 1.26 Control effort and weights for Type 6. The RMS value of the weights shows some lack of smoothness.

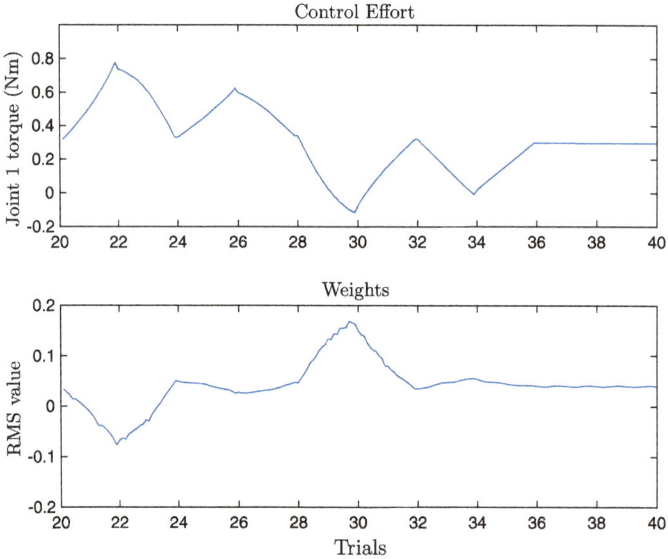

Figure 1.27 Control effort and weights for Type 8. Both control-signal and weight-RMS values demonstrate smoothness.

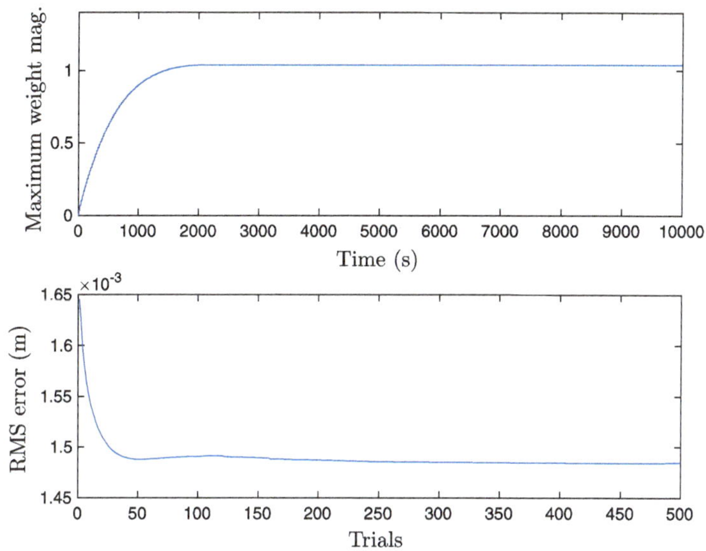

Figure 1.28 The learning curve and weight convergence for Type 6.

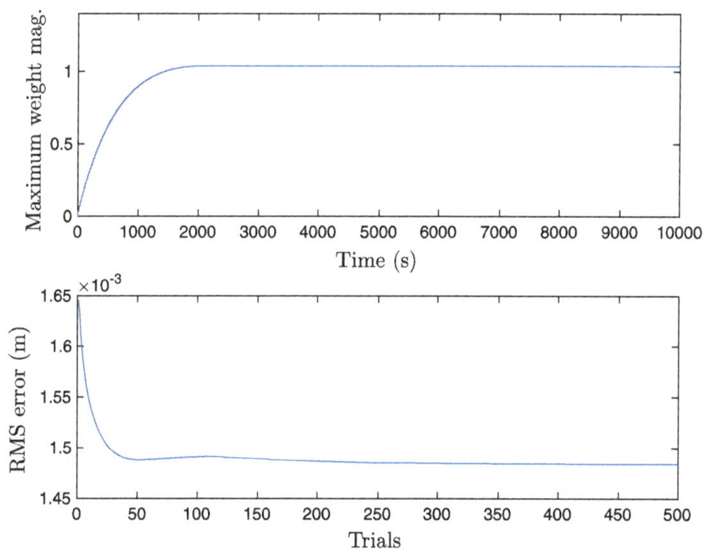

Figure 1.29 The learning curve and weight convergence for Type 8.

a two-link robotic manipulator, as well as significantly outperform a traditional PID+bias control design.

Appendix 1.A

Consider the physical system from (1.5) with the objective of following desired trajectory $x_{\text{des}}(t)$, $\dot{x}_{\text{des}}(t)$, $\ddot{x}_{\text{des}}(t)$, i.e. reducing states errors $e_1 = x - x_{\text{des}}$ and $e_2 = \dot{x} - \dot{x}_{\text{des}}$ to zero. One standard method tries to directly reduce the auxiliary error $z = \Lambda e_1 + e_2$, with positive constant $\Lambda = K_P/K_D$, since $z \equiv 0$ implies a stable system $e_2 = -\Lambda e_1$.

The universal approximation abilities of CMAC imply it can approximate a nonlinear function

$$f(\mathbf{q}) = \mathbf{\Gamma}(\mathbf{q})\mathbf{w} + d(\mathbf{q}), \tag{1.29}$$

where $d(\mathbf{q})$ represents a bounded uniform approximation error. Consider choosing an adaptive-control Lyapunov function

$$V = \frac{1}{2}z^2 + \frac{1}{K_I}\tilde{\mathbf{w}}^{\mathbf{T}}\tilde{\mathbf{w}}, \tag{1.30}$$

where the $\tilde{\mathbf{w}}$ defines the weight errors, the difference between constant ideal weights $\mathbf{w}$ and changing-estimate of the ideal weights used in the network $\hat{\mathbf{w}}$,

$$\tilde{\mathbf{w}} = \mathbf{w} - \hat{\mathbf{w}}. \tag{1.31}$$

Taking the time derivative of the Lyapunov function gives

$$\dot{V} = z\dot{z} - \frac{1}{K_I}\tilde{\mathbf{w}}^{\mathbf{T}}\dot{\hat{\mathbf{w}}} \tag{1.32}$$

where $\dot{z}$ comes from (1.6); note the CMAC neural network can approximate $F(e_1, e_2, x_{des}, \dot{x}_{des}, \ddot{x}_{des})$ so that

$$\begin{aligned} \dot{V} &= z(u + \mathbf{\Gamma}\mathbf{w} + d) - \frac{1}{K_I}\tilde{\mathbf{w}}^{\mathbf{T}}\dot{\hat{\mathbf{w}}}, \\ &= z(u + \mathbf{\Gamma}\hat{\mathbf{w}} + d) + z\mathbf{\Gamma}\tilde{\mathbf{w}} - \frac{1}{K_I}\tilde{\mathbf{w}}^{T}\dot{\hat{\mathbf{w}}}. \end{aligned} \tag{1.33}$$

Substitution of the control law (1.7) and weight-update (1.13) results in

$$\dot{V} = -K_D z^2 + zd - \frac{K_B}{K_I}\tilde{\mathbf{w}}^{T}(\bar{\mathbf{w}} - \tilde{\mathbf{w}}). \tag{1.34}$$

Since the CMAC ensures approximation errors remain bounded, $\|d\| < d_{\max}$,

$$\dot{V} \leqslant -K_{D_{\min}}z^2 + zd_{\max} + \frac{K_B}{K_I}\|\tilde{\mathbf{w}}\|\|\bar{\mathbf{w}}\| + \frac{K_B}{K_I}\|\tilde{\mathbf{w}}\|\|\mathbf{w}\| - \frac{K_B}{K_I}\|\tilde{\mathbf{w}}\|^2. \tag{1.35}$$

Thus $\dot{V} < 0$ when either $|z|$ or $\|\tilde{\mathbf{w}}\|$ becomes large enough, $|z| > \delta_z$ or $\|\tilde{\mathbf{w}}\| > \delta_w$ respectively; thus there exists an elliptical Lyapunov surface on the $(|z|, \|\tilde{\mathbf{w}}\|)$ plane, lying within the region where $\dot{V} < 0$, that denotes a uniform ultimate bound for the signals given by

$$V(z, \tilde{\mathbf{w}}) = V(\delta_z, \delta_w). \tag{1.36}$$

Thus, the nonlinear adaptive control ensures all signals are uniformly ultimately bounded (UUB).

References

[1] C. Macnab, S. Doctolero, The role of unconscious bias in software project failures, in: Intl. Conf. on Software Engineering Research, Management and Applications, Springer Studies in Computational Intelligence 845 (2019) 91–116.

[2] C. Macnab, Developing a computer model of human personalities suitable for robotics and feedback applications, in: IEEE International Conference on Systems, Man and Cybernetics (SMC), IEEE, Bari, Italy, 2019, pp. 3261–3268.

[3] C. Macnab, A feedback model of human personalities, in: IEEE International Conference on Systems Engineering, IEEE, Montreal, 2020, pp. 3261–3268.

[4] C. Macnab, S. Doctolero, Developing a computer model of human personalities suitable for robotics and control applications, in: The Proceedings of the IEEE Conf. on Systems, Man, and Cybernetics, Bari, Italy, 2019, pp. 2907–2914.

[5] R. Kelly, Pd control with desired gravity compensation of robotic manipulators: a review, The International Journal of Robotics Research 16 (1997) 660–672.

[6] H. Ernesto, J.O. Pedro, Iterative learning control with desired gravity compensation under saturation for a robotic machining manipulator, Mathematical Problems in Engineering (2015).

[7] H. Zhang, M. Du, G. Wu, W. Bu, Pd control with rbf neural network gravity compensation for manipulator, Engineering Letters 26 (2018) 236–244.

[8] J. Fujishiro, Y. Fukui, T. Wada, Finite-time pd control of robot manipulators with adaptive gravity compensation, in: IEEE Conference on Control Applications, IEEE, 2016, pp. 898–904.

[9] C. Sun, W. He, W. Ge, C. Chang, Adaptive neural network control of biped robots, IEEE Transactions on Systems, Man and Cybernetics: Systems 47 (2016) 315–326.

[10] Z.-G. Liu, J.-M. Huang, A new adaptive tracking control approach for uncertain flexible joint robot system, International Journal of Automation and Computing 12 (2015) 559–566.

[11] M. Wang, A. Yang, Dynamic learning from adaptive neural control of robot manipulators with prescribed performance, IEEE Transactions on Systems, Man and Cybernetics: Systems 47 (2017) 2244–2255.

[12] S. Dian, Y. Hu, T. Zhao, J. Han, Adaptive backstepping control for flexible-joint manipulator using interval type-2 fuzzy neural network approximator, Nonlinear Dynamics 97 (2019) 1567–1580.

[13] M. Wang, H. Ye, Z. Chen, Neural learning control of flexible joint manipulator with predefined tracking performance and application to Baxter robot, Complexity (2017).

[14] M. Razmi, C. Macnab, Near-optimal neural-network robot control with adaptive gravity compensation, Neurocomputing (2020), https://doi.org/10.1016/j.neucom.2020.01.026.

[15] J. Albus, A new approach to manipulator control: the cerebellar model articulation controller (CMAC), Journal of Dynamic Systems, Measurement, and Control 97 (1975) 220–227.
[16] J. Albus, Data storage in the cerebellar model articulation controller (CMAC), Journal of Dynamic Systems, Measurement, and Control 97 (1975) 228–233.
[17] P. Ioannuou, P. Kokotovic, Instability analysis and improvement of robustness of adaptive control, Automatica 20 (1984) 583–594.
[18] D. Richert, K. Masaud, C. Macnab, Discrete-time weight updates in neural-adaptive control, Soft Computing 17 (2013) 431–444.
[19] C. Naranjo, Character and Neurosis: An Integrative View, Gateways/IDHHB, Nevada City, 1994.
[20] J. Talbott, A Comparison of the Unhealthy Expressions of the Enneagram Types and the Personality Disorders of the DSM-IV-TR, Liberty University, 2017.
[21] D. Lowry, True Colors: Keys to Successful Teaching, True Colors, 1989.
[22] D.W. Merrill, R.H. Reid, Personal Styles & Effective Performance, CRC Press, 1981.
[23] C. Macnab, Biologically-inspired personalities for control systems and robots using nonlinear optimization and feedback theory, in: IEEE International Conference on Cognitive and Computational Aspects of Situation Management, IEEE, Victoria, 2020.
[24] M. Spong, M. Vidyasagar, Robot Dynamics and Control, John Wiley and Sons, New York, 1989.
[25] F. Lewis, S. Jagannathan, A. Yesildirek, Neural Network Control of Robot Manipulators and Nonlinear Systems, Taylor and Francis, Philadelphia, PA, 1999.

CHAPTER 2

Cognitive load estimation for adaptive human–machine system automation

P. Ramakrishnan[a], B. Balasingam[a], and F. Biondi[b]
[a]Department of Electrical and Computer Engineering, University of Windsor, Windsor, ON, Canada
[b]Faculty of Human Kinetics, University of Windsor, Windsor, ON, Canada

Contents

2.1 Introduction

2.1.1 Human–machine automation

Human machine system (HMS) is where a human operator's functioning is integrated to that of a machine. The goal of such a system is to enable the human to effectively operate and control the system, whilst the machine provides feedback information and aids in better performance of the operator. When designing a HMS there are a few factors that need to be considered, such as the extent of automation of the HMS, system design, functions that need to be automated, the level of automation and

Learning Control
https://doi.org/10.1016/B978-0-12-822314-7.00007-9

degree of human involvement in a task controlled by the machine [1]. Traditionally these systems have been explored as binary function allocations where either the machine or the human is assigned to a task. Recently intermediate levels of automation have been developed, where the machine performs the task but the human remains engaged in the task, this evades certain out-of-the-loop performance problems [2]. Adaptive automation has been largely proposed as a method for regulating this out-of-the-loop performance problems, specifically in complex control systems [3]. Attentional factors greatly influence the human interaction with the system. The level of automation in the system could be semi-automated or a fully automated system [4]. In today's world there are several applications for a human machine system: aerospace, aviation [5], e-learning, information processing, autonomous driving, assembly operations [6], manufacturing. The past few decades have seen tremendous progress in system automation. System automation can be broadly classified into four functions—information acquisition, information analysis, decision selection, and action implementation. A system can include all of these functions, or one or more of them at various levels with varied levels of automation [7]. To enhance the overall HMS performance, the task load on the operator must be assigned keeping in mind the fluctuation of psychophysiological functional status of the operator from time to time [8]. In order to better understand the psychophysiological functional status of the operator, it is important to understand the cognitive load experienced by the human performing the task [9].

Cognitive load is a measure of the amount of working memory resources being used. It is a measure of the mental demand experienced by the human while performing a task [10]. The monitoring of cognitive load is useful in enhancing the task performance. As the cognitive task becomes challenging, the amount of working memory required for task completion will gradually increase [11]. Cognitive load measurement plays a vital role in semi-autonomous vehicles, aviation, manufacturing, surgery, mobile learning interfaces to name a few applications [12,13]. Unlike the physical load, it is difficult to measure cognitive load through direct and non invasive techniques. Researchers have suggested that cognitive load can be detected using analytical and empirical methods [9]. Analytical methods are aimed at evaluating the cognitive load by collecting analytical data with methods such as mathematical models and task analysis. Empirical methods involve estimating the cognitive load by collecting subjective data using rating scales, performance data using primary and secondary task techniques

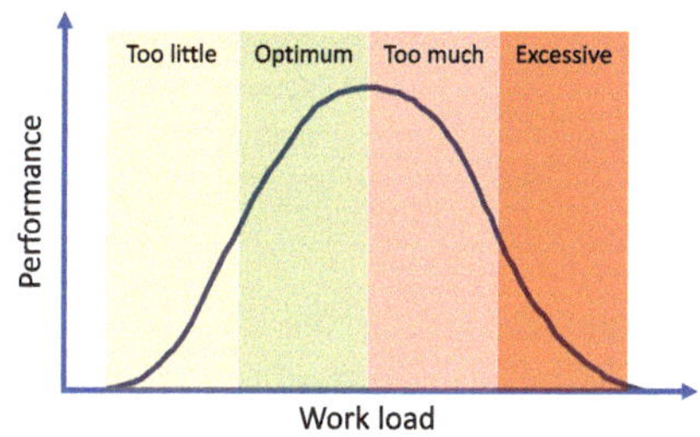

Figure 2.1 Workload and performance. The above curve gives an approximate relationship between workload and performance based on the Yerkes–Dodson Law (YDL).

and physiological data using physiological techniques. These techniques are based on the assumption that changes in cognitive functioning are indicated by physiological variables [14].

The Yerkes–Dodson law (YDL) [15] defines an empirical relationship between arousal and performance as shown in Fig. 2.1. This law dictates that the performance increases with physiological arousal but this is true only up to a certain level. When the level of arousal is too high, the performance decreases, and when the arousal is too low, there is loss of interest, which results in low performance [15]. In order to achieve maximum performance the workload has to be in the optimum range—by providing optimum arousal, maximum performance can be achieved [16]. The YDL can be exploited in many practical situations for performance enhancements: affective computing, logic theory, designing e-learning methods, manufacturing, aerospace, cyber-physical systems to name a few. The level of arousal plays a vital role as it is directly related to the cognitive state of the subject [17]. There exist several methods to identify the optimum level of arousal and predict when the performance is expected to be the highest.

The workload performance curve discussed above has a key role to play in modeling a tool that could be indicative of the perceptual or cognitive workload experienced by the human. This is necessary to help design a better HMS that is able to adapt and re-assign tasks based on the cognitive state of the human performing the task [18]. A generic flow diagram of how this system would function is shown in Fig. 2.2. The prediction of workload plays a vital role in understanding the mental state of the human and varying the arousal/task load. The operational performance of a closed-loop HMS can be enhanced when the machine is able to monitor the operator's cognitive state and adapt accordingly to maximize the ef-

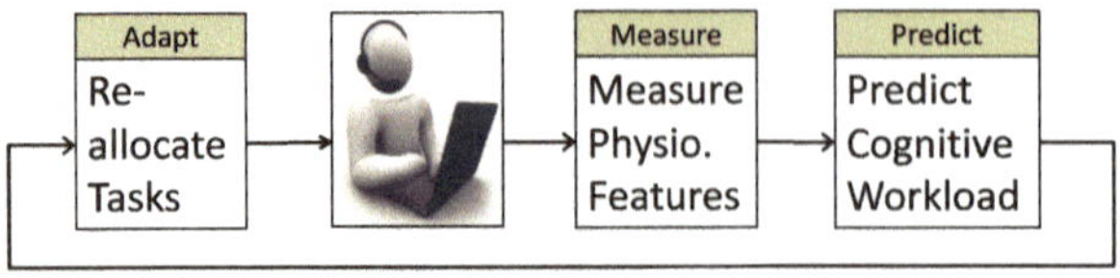

Figure 2.2 Human–machine automation. The goal is to keep the human performance at the optimum level (see Fig. 2.1) by reallocating the tasks.

fectiveness of the HMS [19]. There exist several well-studied techniques of estimating the cognitive workload experienced by the humans; some of these will be discussed in Sect. 2.2.

2.1.2 Cognitive load measures

The basis for designing different levels of human–machine system automation is to understand the cognitive state of the human being [19]. As such, the first step towards designing a better man–machine system is cognitive load estimation—preferably through non-invasive measures. There are several different techniques for cognitive load measurement as shown in Fig. 2.3 which shows a taxonomy of different approaches to it. Broadly these techniques can be classified into objective and subjective measures of cognitive load. Objective approach to cognitive load estimation is further classified as performance based metrics, physiological metrics, and behavioral metrics. Pupil diameter, eye-gaze patterns, eye blinks, heart rate, cardiac activity, electroencephalogram (EEG) activity, event related potential (ERP), and skin conductance are some of the physiological metrics of cognitive workload [20]. Behavioral metrics of cognitive workload include gestures/movements when performing a task that can act as non-verbal outputs of the human body revealing valuable information on how people think and feel [21]. Detection response task (DRT) is a widely accepted standard for a behavioral measure of cognitive load [22]. Considering the task complexity and task workload or stress, the behavior of the human performing the task is said to resort to routine action and reduce the variety in their response to the task as the complexity of the task increases [23]. It would be interesting to understand how committed/engaged the human remains in the task as the task workload increases [24]. Knowledge on how engaged the human is in the task that is under way would help us better understand the cognitive state of the person. The task engagement can be predicted by measuring certain behavioral metrics like their response to visual or tactile stimuli [25]. Pupil diameter (physiological metric) and

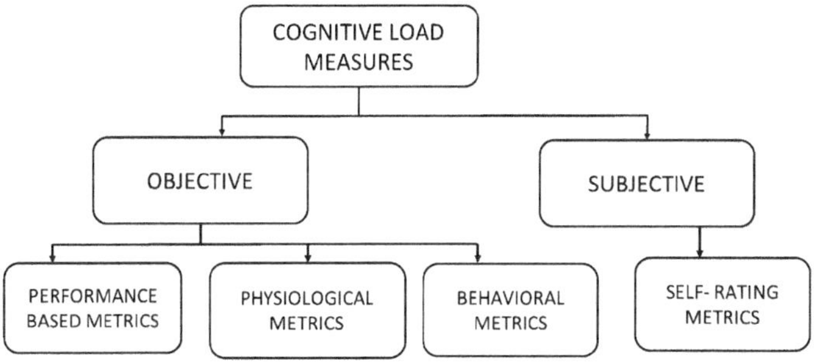

Figure 2.3 Cognitive load measurement approaches. The figure shows the various approaches to cognitive load measurement.

response time (behavioral metric) are two well established and widely used methods for cognitive load estimation [26]. Subjective measure of cognitive load involves a self-rating approach by the person performing the task to rate the task difficulty on various scales. The NASA task load index (NASA-TLX), a retrospective questionnaire, is one of the best known subjective measures of cognitive load [27].

Human physiological signals have been successfully used to indicate the work load during task execution. There exist several techniques to measure the physiological signals some of which are listed in Table 2.1. Although there are several physiological signals that are indicative of the cognitive load, not all of these measures can be reliably used in all applications. Considering the sensitivity, ease of measurement, and usability, many of the current methods listed in Table 2.1 are not ideal for ubiquitous applications. This has given way to the increase in number of wearable devices that can measure the attention given to the task, with physiological signals and be able to constantly estimate the mental load experienced by human subjects [37]. However, it is not possible to use wearable devices to estimate the work load in all applications [38]. In applications where it is not possible to intervene with the activity being performed whilst measuring the cognitive load, researches have led to the use of non-invasive techniques for estimating the same. Applications where invasive techniques (which involve the device being in physical contact with the human subject under study) cannot be deployed, a non-invasive measure of cognitive load have been sought. A few applications where non-invasive measures can be used and how it helps in enhancing the performance of the system are discussed below.

Table 2.1 Physiological metrics of mental workload.

Physiological measurement	Metrics
Event-related brain potentials (ERPs)	Amplitude and latency of P300 component and Early negativities (first 250 milliseconds) [28]
EEG activity	Mean power in α(8-13 Hz), β (14-25 Hz) and θ (4-7 Hz) frequency bands of EEG [29]
MEG activity	Amplitudes of N100m and P200m deflections in the magnetic response to a sensory stimulus, peak latencies of magnetic responses [30]
Brain metabolism	Regional Cerebral Blood Flow (rCBF)(measured by Positron Emission Tomography (PET), functional Magnetic Resonance Imaging (fMRI)), changes in blood oxygenation in dorsolateral prefrontal cortex (measured by fNIR (functional Near Infrared) spectroscopy) [31]
Pupil diameter	Average diameter, Maximum pupil dilation [32]
Eye movements	Saccade frequency and saccade distance, Fixation duration, Dwell patterns, NNI, Entropy [33]
Eye blinks (EOG)	Mean blink duration, blink rate and blink latency [34]
Cardiac activity	Heart rate, heart beat duration, heart rate variability, power in the mid-frequency (0.1 Hz component) and high frequency (vagal tone) ranges of the electrocardiogram (ECG), Interbeat interval (IBI) [35]
Electrodermal activity	Skin conductance/resistance [36]

2.1.3 Some applications

Applications where non-invasive measures need to be deployed include surgery, flight-safety, human-centered design, multi-media learning, to name a few [39]. In driving related environment the non-invasive metrics can be used to estimate the workload experienced by the driver, to enhance the automated features and safety standards [40]. In most of the applications listed non-invasive metrics play a pivotal role in better understanding the cognitive state of the human being.

Technology has grown and diverse teaching techniques have emerged such as e-books, massive online open courses (MOOC), and gamification. Online teaching/learning has become an integral part of education today. Technologies continue to provide revolutionary improvements in traditional learning environments. The pedagogic trend is becoming "student-centered" [41]. The cognitive load theory is an instructional design theory that focuses on designing conditions and principles that enhance learning [42]. The assumptions of this theory are dependent on certain aspects of human cognition. Cognitive load theory assumes that instructional design should address the limitations of working memory which temporarily maintains and stores information [43]. Working memory is the part of human cognition where the central work of multimedia learning takes place [44]. This is due to the fact that working memory processes information before it is stored in long-term memory [45]. Likewise, working memory selects verbal and pictorial information from sensory memory, organizes them, and integrates them [43]. Therefore, one of the basic premises of cognitive load theory is that instructional design for online teaching should optimize cognitive load at a level that does not hinder learning. The point is not to decrease cognitive load but keep it at a level that does not prevent learning [46]. Both cognitive overload and underload do not promote learning. Therefore it is important to maintain optimum load to design better teaching methodologies and techniques. It is also necessary to keep in mind, the way in which working memory functions which will help us design online learning program and keep the end user engaged in the learning task. Pupil dilation can be used as an index of learning [47]. It is one of the important indicators of the learning that takes place remotely.

Most educational institutions use traditional face-to-face instruction. Many online courses are available in which video lectures are used as a medium of instruction [48]. A video lecture may be more complex, paired with slide presentations, interactive quizzes and demonstrations [49]. Online video lectures have become increasingly common in recent years, as evidenced by their use in many organizations, educational institutions, and open learning systems, such as Coursera, Udemy, TED to name some. Video based learning often provides students with additional time to fully understand classroom course materials by allowing them to review lectures repeatedly [50]. Video based learning is targeted at harnessing learning motivation, increases the learning performance, satisfies individual learning needs with varied learning styles, and selects the most appropriate format to facilitate learning [51]. Moreover, attention is said to facilitate the selection of incoming perceptual information and limiting the number of external stimuli processed by the bounded cognitive system to avoid overloading [52]. Importantly a learning process without sustained attention lacks effective identification, learning, and memory [53]. Sustained attention to the content is of priority and concern for effective learning, explaining the need to determine whether different styles of video lectures affect sustained attention in video based scenarios. Research studies have asserted that design of multimedia materials or video lectures should consider the affective state (i.e. a learner's emotional state) [54].

The advent of cloud storage and processing, in combination with the advancements in connectivity, has prompted video-on-demand (VoD) as a new form of television. Unlike traditional video streaming devices like television, in VoD, the viewer browses through the lists of available videos and selects one to play; further, the VoD allows the user to pause and resume videos at any time. The content providers, who earlier had the upper hand in deciding what the viewers will watch and when, are now forced to contest for the attention of the viewers; this prompted a new research area called recommendation systems [55] where the *recommender engines* employ algorithms to suggest videos that the users might be interested to watch. Existing recommendation systems rely on behavioral metrics to estimate the quality of experience (QoE) to suggest videos with an objective to keep the QoE of the user at a higher level. It is interesting to note that aspirational VoD system [56] is not so different in structure compared to the conceptual HMS introduced in Fig. 2.2; what is different though is the fact that the present day VoD systems use behavioral metrics, based on browsing history, to make their predictions. Considering that most video systems are able to connect to physiological sensors of a viewer, such as smart watches and video cameras, the future VoD system will be able to exploit physiological measures to provide an entertainment experience that is optimized to the physical and mental health of the viewer.

With semi-autonomous vehicles rapidly taking over manually controlled vehicles, the important factors contributing to the changeover are the advanced driver assistance systems (ADAS) which aims to facilitate driving with minimal effort by the human driver. The ADAS based intelligent safety systems could improve road safety in terms of crash avoidance, crash severity mitigation and protection, and automatic post-crash notification of collision [57]. ADAS could be useful as an integrated in-vehicle infrastructure based systems which contribute to all of these crash phases. Crash data studies have found that driver error and other human factors contribute to as much as 93 percent of vehicle crashes [58]. Human error can occur due to lack of training on how to use ADAS, the complexity of ADAS, change of human behavior and sole attention on ADAS rather than primary task of driving. Instead of providing the entire control to the driver, ADAS design which can adapt and trigger based on feedback received from the driver's cognitive state would be a better suited system [59]. To detect drowsiness using both detection of driver's face and eyes along with percentage of eye closing data and thereby identifying the state of the eye—the number of blinks can be detected, and depending on the level of drowsiness a warning message can be displayed, or the vehicle can be slowed down and stopped thus avoiding a crash [18]. A method for detecting driver's inattention to facilitate operating mode transitions between driver and driver assistance systems based on cognitive model of human behavior is necessary [60]. Cognitive load delays a driver's response to critical events. Under conditions of high cognitive load, failure of ADAS due to human error can result in undesired incidents such as crashes [61]. The mental process, emotional effort, and motor actions also evoke the pupil dilation. Previous studies have shown that the mean and variance of the pupil dilation increases with cognitive difficulty. It was also seen that eye-tracking can be used for detecting and tracking transient changes in the pupil dilation for multiple levels of cognitive difficulty [62]. Short-duration studies involving pupil dilation suggest that when information is received into the memory pupil dilates slightly, dilation increases when the information is processed and constricts when information is retrieved. For the long-duration task, the peak pupillary dilation is consistently higher than the short duration task but the constriction during memory retrieval is almost similar under the two conditions [63]. Pupil dilation is by far the most trusted non-invasive metric of cognitive load measurement.

Having explained the application of non-invasive measures of cognitive load, the remainder of this chapter is organized as follows: Section 2.2 details several non-invasive measures of cognitive loads; Sect. 2.3 details several experiments where one of these non-invasive measures, the pupil diameter, was experimented with as a measure of cognitive load.

2.2 Noninvasive metrics of cognitive load

In this section, we discuss some non-invasive measures of cognitive load that are possible candidates to the HMS automation concept shown in Fig. 2.2.

2.2.1 Pupil diameter

Time bound averaging of pupil dilation data, along with events that induce higher cognitive load is associated with the central nervous system. When people are given cognitively challenging tasks, the pupil dilates; this phenomenon is known as task-evoked pupillary response [64]. Kahneman

(1973) in his theory of attention concludes that the primary measure of processing loads can be based on out of task evoked pupillary responses [65]. This theory was based on the fact that humans have limited cognitive capacity and this is closely related to the arousal system. Human cognitive processing such as problem solving, language comprehension, is accompanied by pupil dilations. Research shows that any sensory movement—tactile, auditory, or gustatory triggers pupil dilation [66]. Although the dilation in pupil might be small, it adds a lot of predictive strength to the cognitive load detection [62]. The pupillary response has long been known to be directly associated with increased mental activity. Despite the challenges in using pupil dilation as a measure of cognitive load, the recent advancements in the field of eye-tracking have made this a feasible tool for human system automation. Today's technology allows one to obtain pupil diameter measurements using non-invasive, inexpensive cameras [67]. Pupil dilation can also be used as a method of detecting and tracking transient changes with varied levels of cognitive difficulty. The variation of pupil diameter for short duration and long duration tasks are different. For shorter tasks when the information is received into the memory the pupil dilates; the dilation further increases when the information is being processed and finally constricts when the information is retrieved. For long duration tasks the pupil dilation is different [63]. The peak pupillary dilation is consistently higher in tasks with longer duration, but the constriction in the information retrieval phase is almost similar in the two tasks [66].

2.2.2 Eye-gaze patterns

Research on eye-gaze based cognitive load detection dates back to the 18th century when Louis Emile investigated saccadic movement of the eye during a reading task. Changes in the activity of the central nervous system are systematically related to the cognitive processing of the task [68]. Study of eye-gaze patterns not only helps in better understanding of the cognitive load experienced by the person, but it also helps in better designing of billboards, traffic signs, and posters. Eye gaze serves many functions ranging from social, emotional and intellectual. When presented with a cognitively demanding task, when a question is asked, when remembering phone numbers, or while speaking, humans tend to look away from the person they are conversing with, at a distant object, or the sky; this phenomenon is termed gaze-aversion (GA). The GA is also found to increase with the increasing demand of the task and is considered to be a key cue during pedagogical interactions [69]. Often this is considered as a non-verbal indication of the level of concentration of the person on the task being performed. The GA phenomenon typically reflects the need to concentrate on drawing information from memory or engage in spontaneous cognitive processing. Eye-gaze pattern simulation has a slight edge over the pupil diameter based approach of cognitive load detection, which is the fact that it can be used for people with various visual impairments as well [68]. The cognitive load hypothesis related to GA suggests that there is a slight increase in GA when the cue being presented to the person is a visual cue in comparison to any other stimuli [69]. When the need to concentrate on internal cognitive processing increases, we ignore the other cues in front of us by simply looking away from them. This hypothesis is consistent with the finding that GA occurs more in response to objects than faces.

2.2.3 Eye-blink patterns

Eye blinks are said to occur before and after a cognitively demanding task has been performed. Pupil dilation and blink patterns provide complementary and mutually exclusive indices of information processing. Though both these parameters are associated in a way with cognitive load measurement, blinks prevent the measurement of pupil diameter to a certain extent, and mostly these two indices are discussed independently in the literature [34]. However, there are certain studies that show that blinks occur during the early stages of sensory processing and pupil dilation better reflects sustained information processing. An independent literature suggests that blinks are not a random occurrence but blink bursts follow high cognitive load or information processing. Cognitive processing associated with blinks can be sporadic and a combination of these indices gives us a better understanding of the relationship between cognitive load and blink occurrence [70]. However, blink is shown to be a better indicator of perceptual load than other eye tracking metrics [70].

2.2.4 Heart rate

The cognitive load experienced by the driver has always been associated with the driver's heart rate. Unlike any other performance measures or subjective workload assessment, the physiological measure allows a continuous assessment throughout a task [71]. These measures do not necessitate any interruption of the primary task, which is advantageous over studies that purely examine only behavioral responses. Measuring of cardiac activity for assessing cognitive workload can be done by simply measuring the electrical signals that are emitted by the heart. The mental workload also depends on whether the secondary task being performed is visual, auditory or haptic [72]. Cardiac activity measurement offers a great deal of advantage over other devices such as EEG, which require a greater deal of training to be used and are more susceptible to noise from the physical movement of the subject. Heart rate monitors are in comparison cheaper, and they can be used with minimum training by researchers. The most basic of the cardiac assessment feature is the heart rate (HR), analyzed using the magnitude of a particular cardiac subcomponent deflections (commonly referred to as beats per unit of time) or IBI which measures the beat-to-beat variations. Change in cardiac activity is also affected based on factors that vary from one subject to another. The major contributing factors for this variation are the skill, knowledge, and previous experience related to the particular task [35].

2.3 Details of open-loop experiments

This section describes the details of some experiments designed to validate the use of eye-tracking metrics as indicators of cognitive workload.

2.3.1 Unmanned vehicle operators

Autonomous vehicles have seen significant progress in the recent past. It is important to note that the autonomous features are subjected to continuous development; however, available autonomous features can be combined with human operators for increased productivity. Fig. 2.4 illustrates such a scenario where the one-operator-to-one-vehicle is replaced by multiple-operator-missions-to-multiple-vehicles. Unmanned aerial systems (UASs) are good examples of this emerging model [73], which will soon find applications in ground-based autonomous vehicles.

The UASs are becoming ubiquitous in a wide variety of application, such as, military surveillance [73], agriculture [74], and delivery services [75]. Some sophisticated UAS, such as the single camera, 90 minute endurance, RQ-11 Raven and Group 5, 14 hour endurance, MQ-9 Reaper [76], were initially used in defense applications. In recent years, the UAS has found numerous commercial applications.

Fig. 2.5(a) depicts a typical UAS operation where numerous operators need to monitor the progress of the UAS operation, e.g., delivery of an item, applying fertilizer, etc., in real time. Even though each UAV is able to

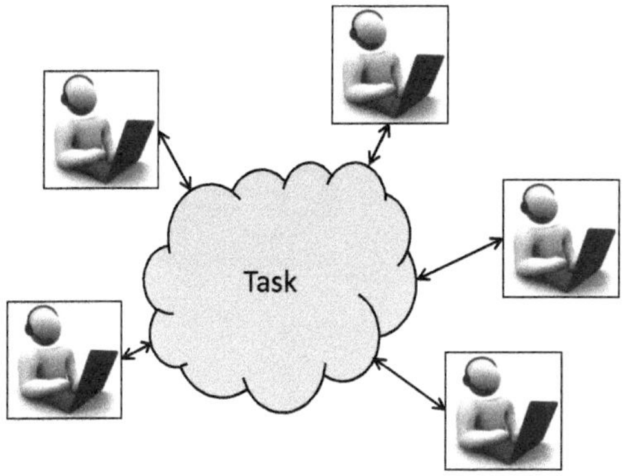

Figure 2.4 Generic illustration of autonomous vehicle operation. This figure illustrates how autonomous vehicle operators will be utilized human operators to execute complex missions.

operate (fly) without a human pilot in it, the overall mission of a UAS requires the continuous attention of human operators. Each UAS is different and the task demand for each operator varies often resulting in sub-optimal tasking and mission performance; it was asserted that 68% of UAS mishaps in the US defense applications were attributable to human error [77,78].

Fig. 2.5(b) depicts the experimental set-up used to simulate an UAS operational scenario [79]. The participant monitors the progress of the UAS mission using the two-screen graphical user interface (GUI) in front of them. During the mission, the UASs would move towards the targets and automatically search the area once within range. Operators were required to respond to events, such as communicating with other operators, updating specific UAS parameters (e.g. altitude), and updating information associated with a target (e.g., location) based upon new information. The mission phase contained three segments, {Block-0, Block-1, Block-2, Block-3} that were designated either of {Easy, Medium, Hard}. The task difficulty was manipulated via the frequency of occurrence of events and the number of new targets added in the segment. The 'Easy' block had events occurring approximately every 75 seconds and 1 new target; the 'Medium' block had events approximately every 45 seconds and 3 new targets; and the 'Hard' block had events approximately every 15 seconds and 4 new targets. The SmartEye system captures eye-tracking at 60 samples/sec while the participant performs tasks of varying difficulty level; each participant completed two approximately twenty minute sessions. Fig. 2.5(c) shows the pupil dilation data collected from one participant during differ-

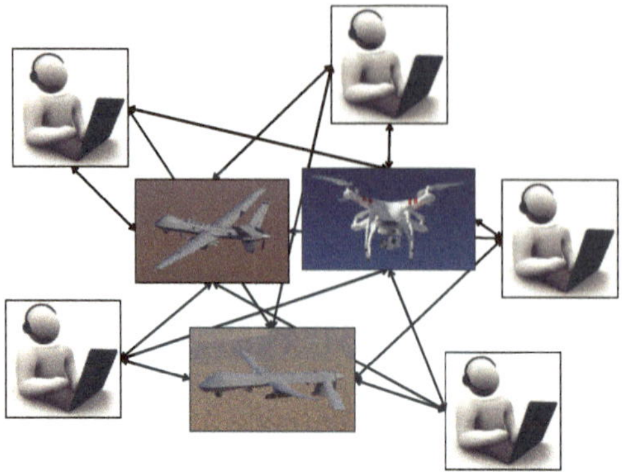

(a) Conceptual unmanned aerial systems (UAS) operation

(b) Measurement set-up from one participant

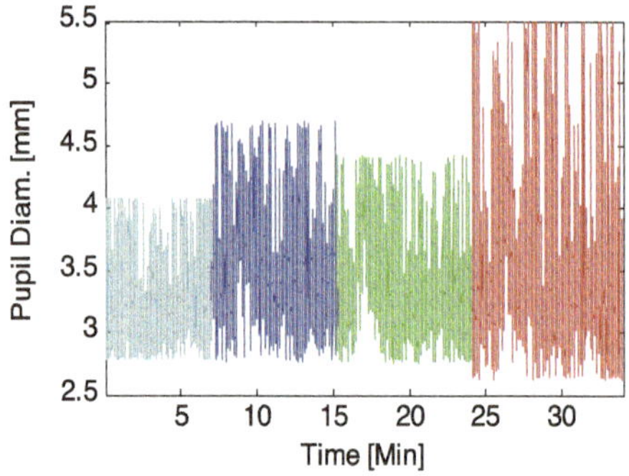

(c) Pupil diameter during tasks of varying difficulty level [80]

Figure 2.5 Simulated UAS management task.

ent blocks. Statistical analysis of this data (collected from 23 participants) showed some level of success in detecting different levels of cognitive load experienced by participants using pupil diameter measurements [79].

In summary, the experimental data collection designed around a simulated UAS operation showed that the cognitive load experienced by individual while engaging in real-world-like activities can be detected using non-invasive measures, such a pupil diameter. It must be kept in mind that much more research needs to be done in order to develop a *closed-loop HMS automation* as illustrated in 2.2.

2.3.2 Memory recall tasks

Memory recall tasks are useful to emulate tasks requiring varying levels of cognitive load, such as, operating equipment, driving, and learning. Fig. 2.6(a) shows the interface used to conduct a digit span experiment in [80]. A brief description of the experimental procedure is as follows: The participant is given digits of sizes 3, 5, 7 and 9 under three different screen luminance conditions (black, gray and white) in audio format; the participants were told to focus on a central fixation cross (a "+" sign ~50 pixels tall and wide) that was offset from the background color (80 brighter for the black and gray backgrounds, and 80 darker for the white background) while he/she was listening to the audio. The string of numbers was then sequentially presented at a rate of 1 digit per second. Once the number is given, a numeric keypad, similar to the one shown in Fig. 2.6(a), appeared on the screen and the participant is asked to use the mouse to input the string of numbers ("2, 6, 1, 8, 4") that they heard by clicking on the corresponding digit in the order they heard. The keypad ensures that participants continued to fixate on the screen while they were making a response; i.e., a verbal response from the participant did not require them to focus on the screen while responding. The keypad had a "back" button allowing the participants to change their response; when satisfied, the participants clicked the submit button.

Fig. 2.6(b) shows the pupil diameter (in pixels) measured by the Gazepoint GP3 eye tracker [81]. It was reported in [80] that the pupil diameter had a positive correlation with the difficulty for a given background color. However, the pupil diameter can be seen to be more sensitive to the background color than to the cognitive difficulty—this observation underlines the challenges involved in using pupil diameter as a sole indicator of cognitive load. This observation emphasizes the need to study other metrics, such as eye-gaze patterns, eye-blink patterns, hear rate related metrics, all of which can be measured through relatively non-invasive means.

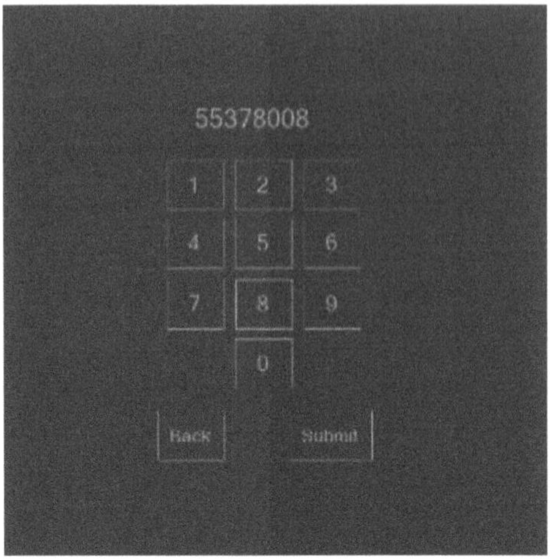

(a) Digit-span task interface (gray background)

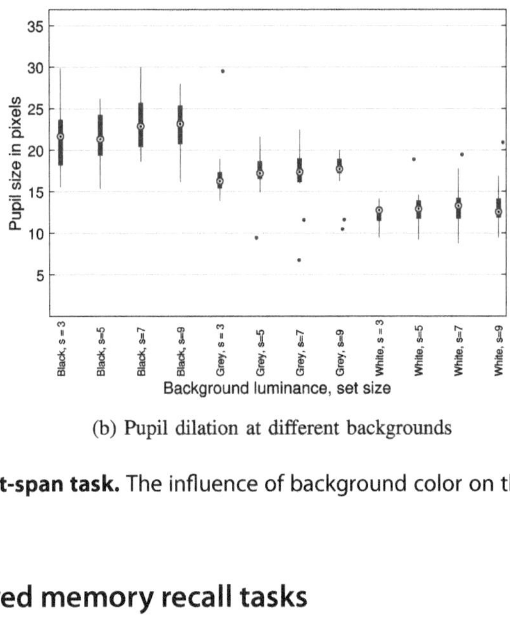

(b) Pupil dilation at different backgrounds

Figure 2.6 Digit-span task. The influence of background color on the pupil diameter is shown.

2.3.3 Delayed memory recall tasks

Delayed memory recall tasks or the *n-back task* requires participants to decide whether each stimulus in a sequence matches the one that appeared n items ago. Although n-back has become a standard executive working memory (WM) measure in cognitive neuroscience [82]. N back task has a serial presentation of stimulus in the form of audio spaced few seconds apart and the participant had to repeat out loud number they heard. This task is further classified into three stages of increasing difficulty. First one is *zero-back* in this task participant had to repeat same number that they just heard. Next one is called *one-back* where the participant had to repeat one number previous to the number they just heard. Similarly, in the *two-back* task, the participant had to repeat the number they heard two numbers prior. The difficulty of the n-back tasks increase with the delay, i.e., a 2-back task is more cognitively demanding than a 1-back task which is more difficult than a 0-back one. Refer to Fig. 2.7 for a sample stimulus (usually delivered in audio format) and the expected response (collected through voice recordings) during 0-back, 1-back and 2-back tasks.

Fig. 2.7(a) shows an experimental setup designed to analyze the pupil diameter in response to n-back tasks of three different difficulty levels. The Gaze point GP3 [81] eye tracker was used to collecting the eye tracking data, response time was collected using detection response task. Entire experiment was divided into two stages: stage 1 had four conditions, and stage 2 had 3 conditions. The participants were asked to look at a '+' sign on the monitor (see Fig. 2.7(a)) during the experiment. The first stage (dual-experiment) of the experiment consisted of four conditions namely: Control, 0-back, 1-back and 2-back. Data collected from the participants involved reaction time, Pupil diameter, eye gaze, eye blinks, NASA task load index form and the n-back response. The second stage (single-experiment) of the experiment consisted of four conditions namely: 0-back, 1-back and 2-back. Data collected from the participants involved pupil diameter, eye gaze, eye blinks, NASA task load index form and the n-back response. The mean pupil diameter data collected from 28 participants during the dual-experiment in Fig. 2.7; here, an increasing trend can be observed for mean pupil diameter with the n-back difficulty.

2.3.4 Simulated driving

This subsection explains a study where n-back tasks are used as distractors during a driving in order to understand the feasibility of estimating cognitive difficulty level based on pupil diameter [83]. In order to simulate driving, a Logitech G29 driving wheel was used along with a driving simulator that allowed to simulate various driving scenarios. The experimental set up is shown in Fig. 2.8(a). The task variation for every participant was Baseline ECG, Control, 0-Back and 2-Back. A Latin square was used to counterbalance the experiment. The baseline task involved the collection of only ECG data when the participants were not performing any task, but were asked to look at the screen with a black cross in the middle of the screen. The control task involved the participant to perform the DRT

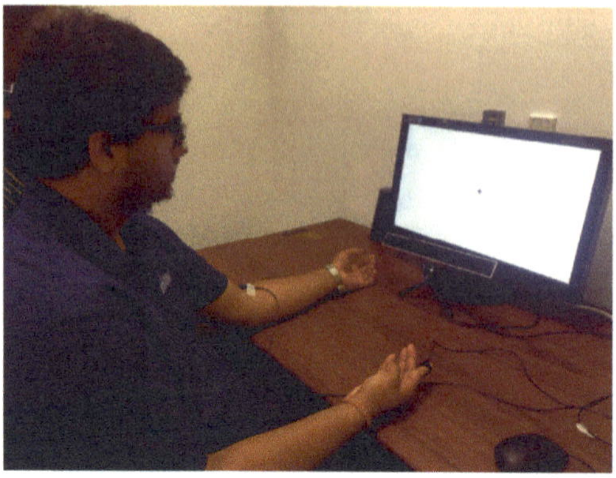

(a) Experimental setup

stimulus	0-back	1-back	2-back
8	8	-	-
7	7	8	-
4	4	7	8
5	5	4	7
2	2	5	4
3	3	2	5
1	1	3	2
9	9	1	3
6	6	9	1
0	0	6	9

(b) Stimulus and expected response

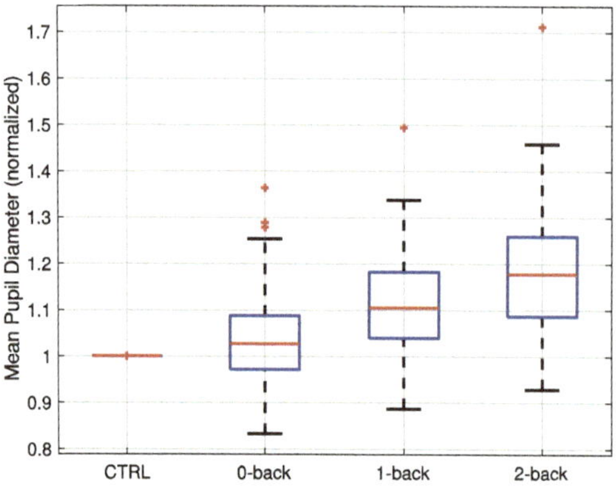

(c) Box plot

Figure 2.7 Delayed memory recall task. Stimulus delivered as an audio play back and the participant is asked to speak the response. The cognitive difficulty increases with the delay.

(a) Experimental setup

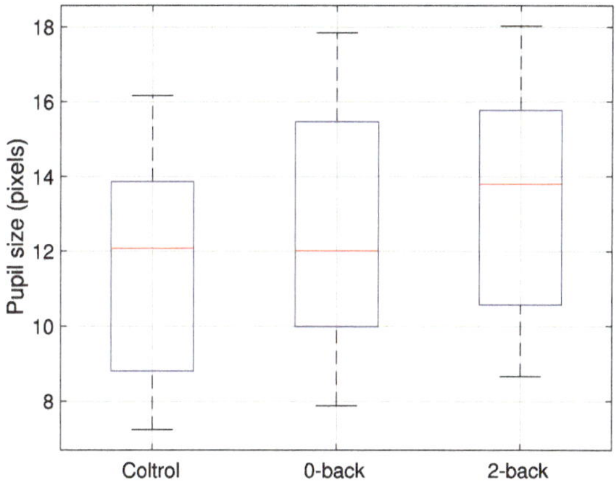

(b)

Figure 2.8 Simulated driving task.

task while driving on the simulator. The response time, pupil diameter, eye-gaze, eye-blinks were recorded. In the 0-back and 2-back tasks, the participants were required to perform the driving task, DRT, and respond to the n-back audio. Reaction time, pupil diameter, eye gaze, eye blinks, ECG and NASA task load index, n-back audio response were recorded for the 0-back and 2-back tasks. Each of the conditions took around 5 minutes to complete. The mean pupil diameter data collected from 16 participants is represented as a box plot in Fig. 2.8(b). Once again, an increasing trend in the mean pupil diameter can be observed from this figure with increasing levels of distraction in drivers.

2.4 Conclusions and discussions

This chapter provides some insights into the current progress in human–machine system automation using physiological metrics—particularly pupil diameter—as a measure of cognitive load experienced by humans. Details of four experiments conducted to observe the changes in pupil diameter in response to mental workloads that were simulated to represent realistic human–machine interactive activities were presented in this chapter. The conclusion is that the pupil diameter is a useful indicator of cognitive load; however, it is affected by other stimuli, such as the ambient light. Hence, it is important to investigate other non-invasive metrics, such as eye-gaze patterns, eye-blink patterns and cardiovascular metrics, in order to obtain robust information about the cognitive load experienced by humans.

2.5 List of abbreviations

ADAS	Advanced driver assistance systems
DRT	Detection response task
ECG	Electrocardiogram
EEG	Electroencephalogram
EOG	Electrooculogram
ERP	Event related potential
fNIR	Functional near infrared spectroscopy
GA	Gaze aversion
GP	Gaze point
HMS	Human machine system
HR	Heart rate
IBI	Inter beat interval
MEG	Magnetoencephalography
MOOC	Massive online open courses
NASA	National Aeronautics and Space Admin.
PET	Positron emission tomography
QoE	Quality of experience
TLX	Task load index
VoD	Video on demand
UAS	Unmanned aerial system
WM	Working memory
YDL	Yerkes–Dodson law

References

[1] D. Kaber, M. Wright, L. Prinzel, M. Clamann, Adaptive automation of human–machine system information-processing functions, Human Factors: The Journal of Human Factors and Ergonomics Society 47 (4) (Jan 2005) 730–741.
[2] M.R. Endsley, Level of automation effects on performance, situation awareness and workload in a dynamic control task, Ergonomics 42 (3) (1999) 462–492, PMID: 10048306.
[3] D.B. Kaber, J.M. Riley, Adaptive automation of a dynamic control task based on secondary task workload measurement, International Journal of Cognitive Ergonomics 3 (3) (1999) 169–187.
[4] R. Parasuraman, Human–computer monitoring, Human Factors 29 (6) (1987) 695–706.
[5] Y. Lim, S. Ramasamy, A. Gardi, T. Kistan, R. Sabatini, Cognitive human–machine interfaces and interactions for unmanned aircraft, Journal of Intelligent and Robotic Systems: Theory and Applications 91 (3–4) (2018) 755–774.
[6] G. Andrianakos, N. Dimitropoulos, G. Michalos, S. Makris, An approach for monitoring the execution of human based assembly operations using machine learning, Procedia CIRP 86 (2020) 198–203, Elsevier.
[7] T.B. Sheridan, Humans and Automation: System Design and Research Issues, Human Factors and Ergonomics Society, 2002.
[8] J.-H. Zhang, J.-J. Xia, J.M. Garibaldi, P.P. Groumpos, R.-B. Wang, Modeling and control of operator functional state in a unified framework of fuzzy inference Petri nets, Computer Methods and Programs in Biomedicine 144 (2017) 147–163.
[9] B. Xie, G. Salvendy, Prediction of mental workload in single and multiple tasks environments, International Journal of Cognitive Ergonomics 4 (3) (2000) 213–242.
[10] F.G.W.C. Paas, J.J.G. van Merriënboer, J.J. Adam, Measurement of cognitive load in instructional research, Perceptual and Motor Skills 79 (1) (1994) 419–430, PMID: 7808878.
[11] T.F. Yap, J. Epps, E. Ambikairajah, E.H.C. Choi, Formant frequencies under cognitive load: effects and classification, EURASIP Journal on Advances in Signal Processing 30 (2011).
[12] R. Deegan, Mobile learning application interfaces: First steps to a cognitive load aware system, International Association for Development of the Information Society, 2013, pp. 101–108.
[13] I.O. for Standardization, Road vehicles—transport information and control systems—detection–response task (drt) for assessing attentional effects of cognitive load in driving, 2016.
[14] F. Paas, J.E. Tuovinen, H. Tabbers, P.W. Van Gerven, Cognitive load measurement as a means to advance cognitive load theory, Educational Psychologist 38 (1) (2003) 63–71.
[15] E.J. Jeong, F.A. Biocca, Are there optimal levels of arousal to memory? Effects of arousal, centrality, and familiarity on brand memory in video games, Computers in Human Behavior 28 (2) (2012) 285–291.
[16] A. Welford, Stress and performance, Ergonomics 16 (5) (1973) 567–580.
[17] W. Dongrui, G. Courtney, B.J. Lance, S.S. Narayanan, M.E. Dawson, K.S. Oie, T.D. Parsons, Optimal arousal identification and classification for affective computing using physiological signals: virtual reality Stroop task 1 (Jul 2010) 109–118.
[18] T. Hong, H. Qin, Drivers drowsiness detection in embedded system, in: 2007 IEEE International Conference on Vehicular Electronics and Safety, IEEE, 2007, pp. 1–5.
[19] N. Pongsakornsathien, Y. Lim, A. Gardi, S. Hilton, L. Planke, R. Sabatini, T. Kistan, N. Ezer, Sensor networks for aerospace human–machine systems, Sensors 19 (2019), Multidisciplinary Digital Publishing Institute.

[20] T. Van Gog, L. Kester, F. Nievelstein, B. Giesbers, F. Paas, Uncovering cognitive processes: different techniques that can contribute to cognitive load research and instruction, Computers in Human Behavior 25 (2) (2009) 325–331.
[21] T.F. Yap, J. Epps, E. Ambikairajah, E.H.C. Choi, Detecting users' cognitive load by galvanic skin response with affective interference, ACM Transactions on Interactive Intelligent Systems 7 (2017) 1–20, ACM New York, NY, USA.
[22] S. Lee, H. Shin, C. Hahm, Effective ppg sensor placement for reflected red and green light and infrared wristband-type photo-plethysmography, 2016, pp. 556–558.
[23] J. Fransoo, V. Weirs, Action variety of planners: cognitive load and requisite variety, Journal of Operations Management 24 (Dec 2006) 813–821, Elsevier.
[24] E.J. Lawler, J. Yoon, Commitment in exchange relations: test of a theory of relational cohesion, American Sociological Review 61 (1996) 89–108, JSTOR, American Sociological Association, Sage Publications, Inc.
[25] T. Cegovnik, K. Stojmenova, G. Jakus, J. Sodnik, An analysis of the suitability of a low cost eye tracker for assessing cognitive load of drivers, Applied Ergonomics 68 (2018) 1–11, Elsevier.
[26] M. Kutila, M. Jokela, T. Mäkinen, J. Viitanen, G. Markkula, T. Victor, Driver cognitive distraction detection: feature estimation and implementation, Proceedings of the Institution of Mechanical Engineers, Part D: Journal of Automobile Engineering 221 (9) (2007) 1027–1040.
[27] S.G. Hart, Nasa-task load index (nasa-tlx): 20 years later, in: Proceedings of the Human Factors and Ergonomics Society Annual Meeting, Vol. 50, Sage Publications Sage CA, Los Angeles, CA, 2006, pp. 904–908.
[28] A.-M. Brouwer, M.A. Hogervorst, J.B.F. Van Erp, T. Heffelaar, P.H. Zimmerman, R. Oostenveld, Estimating workload using eeg spectral power and erps in the n-back task, Journal of Neural Engineering 9 (4) (2012).
[29] A. Gevins, M.E. Smith, H. Leong, L. McEvoy, S. Whitfield, R. Du, G. Rush, Monitoring working memory load during computer-based tasks with eeg pattern recognition methods, Human Factors 40 (1) (1998) 79–91.
[30] M. Stenroos, J. Sarvas, Bioelectromagnetic forward problem: isolated source approach revis(it)ed, Physics in Medicine and Biology 57 (11) (2012) 3517–3535.
[31] E. Molteni, D. Contini, M. Caffini, G. Baselli, L. Spinelli, R. Cubeddu, S. Cerutti, A.M. Bianchi, A. Torricelli, Load-dependent brain activation assessed by time-domain functional near-infrared spectroscopy during a working memory task with graded levels of difficulty, Journal of Biomedical Optics 17 (5) (2012).
[32] J. Hyönä, J. Tommola, A.-M. Alaja, Pupil dilation as a measure of processing load in simultaneous interpretation and other language tasks, The Quarterly Journal of Experimental Psychology 48 (3) (1995) 598–612.
[33] J. Gaspar, C. Carney, The effect of partial automation on driver attention: a naturalistic driving study, Human Factors 61 (8) (2019) 1261–1276, PMID: 30920852.
[34] G. Seigle, N. Ichikawa, S. Steinhauer, Blink before you and after you think: blinks occur prior to and following cognitive load indexed by pupillary changes, Psychophysiology 45 (2008) 679–687.
[35] A.M. Hughes, G.M. Hancock, S.L. Marlow, K. Stowers, E. Salas, Cardiac measures of cognitive workload: a meta-analysis, Human Factors: The Journal of the Human Factors and Ergonomics Society 61 (3) (2019) 393–414.
[36] C.-J. Chao, S.-Y. Wu, Y.-J. Yau, W.-Y. Feng, F.-Y. Tseng, Effects of three-dimensional virtual reality and traditional training methods on mental workload and training performance, Human Factors and Ergonomics in Manufacturing 27 (4) (2017) 187–196, Wiley Online Library.
[37] Y.-M. Huang, Y.-P. Cheng, S.-C. Cheng, Y.-Y. Chen, Exploring the correlation between attention and cognitive load through association rule mining by using a brainwave sensing headband, IEEE Access 8 (2020) 38880–38891.

[38] C. Saitis, M.Z. Parvez, K. Kalimeri, Cognitive load assessment from eeg and peripheral biosignals for the design of visually impaired mobility aids, Wireless Communications and Mobile Computing 2018 (2018).

[39] K. Krejtz, A.T. Duchowski, A. Niedzielska, C. Biele, I. Krejtz, Eye tracking cognitive load using pupil diameter and microsaccades with fixed gaze, PLoS ONE 13 (9) (2018).

[40] D.L. Strayer, J. Turrill, J.M. Cooper, J.R. Coleman, N. Medeiros-Ward, F. Biondi, Assessing cognitive distraction in the automobile, Human Factors 57 (8) (2015) 1300–1324.

[41] Z. Nie, Y. Lu, Design on Internet based information management system for digital protection, Dianli Xitong Zidonghua/Automation of Electric Power Systems 24 (20) (2000) 45–48.

[42] J. Sweller, Working memory, long-term memory, and instructional design, Journal of Applied Research in Memory and Cognition 5 (4) (2016) 360–367.

[43] A. Baddeley, Working memory: looking back and looking forward, Nature Reviews Neuroscience 4 (2003) 829–839.

[44] C.I. Johnson, R.E. Mayer, A testing effect with multimedia learning, Journal of Educational Psychology 101 (3) (2009).

[45] A. Baddeley, Working memory 255 (5044) (1992) 556–559.

[46] K. Kozan, The incremental predictive validity of teaching, cognitive and social presence on cognitive load, Internet and Higher Education 31 (2016) 11–19.

[47] C. Sibley, J. Coyne, C. Baldwin, Pupil dilation as an index of learning, in: Proceedings of the Human Factors and Ergonomics Society Annual Meeting, vol. 55, SAGE Publications Sage CA, Los Angeles, CA, 2011, pp. 237–241.

[48] C.-M. Chen, C.-H. Wu, Effects of different video lecture types on sustained attention, emotion, cognitive load, and learning performance, Computers and Education 80 (2015) 108–121.

[49] D.S. Osborn, Using video lectures to teach a graduate career development course, 2010.

[50] H. Brecht, S. Ogilby, Enabling a comprehensive teaching strategy: video lectures, Journal of Information Technology Education: Innovations in Practice 7 (1) (2008) 71–86.

[51] K. Hornbæk, D.T. Engberg, J. Gomme, Video lectures: Hci and e-learning challenges, in: Workshop on Human–Computer Interaction and e-Learning, 2002.

[52] J. Driver, A selective review of selective attention research from the past century, British Journal of Psychology 92 (1) (2001) 53–78.

[53] J. Lachter, K.I. Forster, E. Ruthruff, Forty-five years after broadbent (1958): still no identification without attention, Psychological Review 111 (4) (2004) 880.

[54] Z. Wang, L. Sun, X. Chen, W. Zhu, J. Liu, M. Chen, S. Yang, Propagation-based social-aware replication for social video contents, in: Proceedings of the 20th ACM International Conference on Multimedia, 2012, pp. 29–38.

[55] J. Davidson, B. Liebald, J. Liu, P. Nandy, T. Van Vleet, U. Gargi, S. Gupta, Y. He, M. Lambert, B. Livingston, et al., The youtube video recommendation system, in: Proceedings of the Fourth ACM Conference on Recommender Systems, 2010, pp. 293–296.

[56] D.R. Pasupuleti, Cognitive video streaming, 2015.

[57] A. Shaout, D. Colella, S. Awad, Advanced driver assistance systems-past, present and future, in: International Computer Engineering Conference, 2011, pp. 72–82.

[58] A. Beck, K.J. Kek, Pupillometry and sensor fusion for monitoring and predicting a vehicle operator's condition, Feb. 7 (2019), US Patent App. 15/666,017.

[59] E. Bekiaris, A. Stevens, Common risk assessment methodology for advanced driver assistance systems, Transport Reviews 25 (3) (2005) 283–292.

[60] N. Kuge, T. Yamamura, O. Shimoyama, A. Liu, A driver behavior recognition method based on a driver model framework, tech. rep., SAE Technical Paper, 2000.

[61] N. Lavie, Attention, distraction, and cognitive control under load, Current Directions in Psychological Science 19 (3) (2010) 143–148.
[62] P. Mannaru, B. Balasingam, K. Pattipati, C. Sibley, J. Coyne, Cognitive context detection for adaptive automation.
[63] J. Beatty, D. Kahneman, Pupillary changes in two memory tasks, Psychonomic Science 5 (10) (1966) 371–372.
[64] J. Beatty, Task-evoked pupillary responses, processing load, and the structure of processing resources, Psychological Bulletin 91 (2) (1982) 276.
[65] D. Kahneman, Attention and effort, vol. 1063, Citeseer, 1973.
[66] B.L.-W.J. Beatty, The pupillary system, in: Handbook of Psychophysiology, vol. 2, 2000, pp. 142–162.
[67] S. Petridis, T. Giannakopoulos, C.D. Spyropoulos, Unobtrusive low cost pupil size measurement using webcameras, 2013.
[68] N.J. Wade, Pioneers of eye movement research, 2010.
[69] D.G. Sneddon, F. Phelps, Gaze aversion: a response to cognitive or social difficulty?, Memory and Cognition 33 (4) (2005) 727–733.
[70] P. Biswas, P. Langdon, Multimodal intelligent eye-gaze tracking system, International Journal of Human–Computer Interaction 31 (4) (2015) 277–294.
[71] T.-H.-H. Dang, A. Tapus, Physiological signals in driving scenario: how heart rate and skin conductance reveal different aspects of driver's cognitive load, in: International Conference on Physiological Computing Systems, pp. 378–384.
[72] C. Wickens, Multiple resources and mental workload, Human Factors 50 (2008) 449–455.
[73] FY 2009-2034 Unmanned Systems Integrated Roadmap, tech. rep, Department of Defense, Washington, DC, 2009.
[74] H. Xiang, L. Tian, Development of a low-cost agricultural remote sensing system based on an autonomous unmanned aerial vehicle (uav), Biosystems Engineering 108 (2) (2011) 174–190.
[75] A. Goodchild, J. Toy, Delivery by drone: an evaluation of unmanned aerial vehicle technology in reducing CO_2 emissions in the delivery service industry, Transportation Research Part D, Transport and Environment 61 (2018) 58–67.
[76] J. Gertler, Us unmanned aerial systems, DTIC Document, 2012.
[77] Report to Congress on Future Unmanned Aircraft Systems Training, Operations, and Sustainability, tech. rep, US Department of Defense, 2012.
[78] K.W. Williams, A summary of unmanned aircraft accident/incident data: human factors implications, tech. rep., DTIC Document, 2004.
[79] P. Mannaru, B. Balasingam, K. Pattipati, C. Sibley, J. Coyne, Cognitive Context Detection Using Pupillary Measurements, Next-Generation Analyst IV, vol. 9851, International Society for Optics and Photonics, 2016, p. 98510Q.
[80] P. Mannaru, B. Balasingam, K. Pattipati, C. Sibley, J.T. Coyne, Performance evaluation of the gazepoint gp3 eye tracking device based on pupil dilation, in: International Conference on Augmented Cognition, Springer, 2017, pp. 166–175.
[81] gazept, gazepoint eye tracker website, https://www.gazept.com/. (Accessed 20 November 2018).
[82] M.J. Kane, A.R. Conway, T.K. Miura, G.J. Colflesh, Working memory, attention control, and the n-back task: a question of construct validity, Journal of Experimental Psychology: Learning, Memory, and Cognition 33 (3) (2007) 615.
[83] T. Cegovnik, K. Stojmenova, G. Jakus, J. Sodnik, An analysis of the suitability of a low-cost eye-tracker for assessing the cognitive load of drivers, Applied Ergonomics 68 (2018) 1–11.

CHAPTER 3

Comprehensive error analysis beyond system innovations in Kalman filtering

Jianguo Wang, Aaron Boda, and Baoxin Hu
Department of Earth and Space Science and Engineering, Lassonde School of Engineering, York University, Toronto, ON, Canada

Contents

3.1 Introduction

As Sorenson [24] pointed out, the developments beginning with Wiener's work and culminating with Kalman's reflect fundamentally the changes that have occurred in control systems theory. In a Kalman filter, the state-space approach is used to deal with the system that gives rise to the observed output for a kind of dynamic system. Since the launch of GPS project in the 1970s and with the maturity and popularization of modern kinematic positioning techniques such as GNSS, inertial navigation, cameras, LiDAR during the last few decades, the Kalman filter absolutely has become the dominant tool to optimally estimate kinematic and/or dynamic states, e.g., position, velocity, acceleration and platform attitude of a moving platform, as the least squares method has been in traditional parameter estimation.

Learning Control
https://doi.org/10.1016/B978-0-12-822314-7.00008-0

However, it has not been widely implemented as standardized as the least squares method, especially in terms of comprehensive error analysis. Apparently, comprehensive error analysis is of paramount importance in a Kalman filter when it comes to the positional accuracy level from meters to centimeters required in plenty of applications, to name a couple: RTK GNSS, and multisensor integrated kinematic positioning and navigation.

This manuscript systematically overviews and outlines the theoretical aspects and practical execution of comprehensive error analysis in Kalman filtering. A summary of the standard Kalman filtering model in discrete time is given in Section 3.2. Section 3.3 reviews two alternate formulations of a Kalman filtering algorithm in order to exploit different observations of error sources. Three essential constituents of comprehensive error analysis in Kalman filtering are discussed: redundancy contribution (Section 3.4), variance of unit weight and variance component estimation (Section 3.5) and test statistics (Section 3.6). Further, a numerical example from a real land vehicle for multisensor integrated kinematic positioning and navigation is given and analyzed in Section 3.7. The manuscript ends with Section 3.8, Remarks.

3.2 Standard formulation of Kalman filter after minimum variance principle

In general, the Kalman filter consists of system and measurement models to provide a recursive formulation to estimate the state vector by minimizing its mean squared errors, i.e., after the minimum variance principle [8]. This section summarizes the standard form of a Kalman filter.

Let us consider a linear or linearized system described in state space and the data are made available over a discrete time series $t_0, t_1, ..., t_k, ..., t_N$. Each time instant corresponds to an observation epoch and is described by $0, 1, ..., k, ..., N$ for simplification. For the same reason, this manuscript omits the deterministic system input vector without loss of generality.

At an arbitrary observation epoch k $(1 \leq k \leq N)$, the system and measurement models are given as follows:

$$\mathbf{x}(k) = \mathbf{A}(k, k-1)\mathbf{x}(k-1) + \mathbf{B}(k, k-1)\mathbf{w}(k) \tag{3.1}$$

$$\mathbf{z}(k) = \mathbf{C}(k)\mathbf{x}(k) + \mathbf{\Delta}(k) \tag{3.2}$$

where $\mathbf{x}(k)$ is the n-dimensional state-vector, $\mathbf{z}(k)$ is the p-dimensional observation vector, $\mathbf{w}(k)$ is the m-dimensional process noise vector, $\Delta(k)$ is the p-dimensional measurement noise vector, $\mathbf{A}(k, k-1)$ is the $n \times n$ coefficient matrix of $\mathbf{x}(k)$, $\mathbf{B}(k, k-1)$ is the $n \times m$ coefficient matrix of $\mathbf{w}(k)$, $\mathbf{C}(k)$ is the $p \times n$ coefficient matrix of $\mathbf{z}(k)$. Furthermore, it is assumed that $\mathbf{w}(k)$ and $\Delta(k)$ conform to

$$\mathbf{w}(k) \sim N(\mathbf{o}, \mathbf{Q}(k)) \tag{3.3}$$

$$\Delta(k) \sim N(\mathbf{o}, \mathbf{R}(k)) \tag{3.4}$$

with their zero means and positive definite variance matrices $\mathbf{Q}(k)$ and $\mathbf{R}(k)$, respectively, where $N(a, b)$ represents a normal distribution with a and b as its expectation (vector) and variance (matrix). Between two different observation epochs i and j $(i \neq j)$, they satisfy

$$Cov(\mathbf{w}(i), \mathbf{w}(j)) = \mathbf{O} \tag{3.5}$$

$$Cov(\Delta(i), \Delta(j)) = \mathbf{O} \tag{3.6}$$

$$Cov(\mathbf{w}(i), \Delta(j)) = \mathbf{O} \tag{3.7}$$

Generally, the initial state vector is given as $\mathbf{x}(0)$ with the associated variance matrix $\mathbf{D}_{xx}(0)$, which is assumed to be independent of $\mathbf{w}(k)$ and $\Delta(k)$ for all k:

$$Cov(\mathbf{w}(k), \mathbf{x}(0)) = \mathbf{O} \tag{3.8}$$

$$Cov(\Delta(k), \mathbf{x}(0)) = \mathbf{O} \tag{3.9}$$

After the minimum variance principle under the assumption of an unbiased estimate of $\mathbf{x}(k)$,

$$\left.\begin{array}{l} E\{\hat{\mathbf{x}}(k)\} = \mathbf{x}(k) \\ E\{[\mathbf{x}(k) - \hat{\mathbf{x}}(k)][\mathbf{x}(k) - \hat{\mathbf{x}}(k)]^T\} = \mathbf{min} \end{array}\right\} \tag{3.10}$$

one derives the Kalman filtering algorithm at epoch k from epoch $k-1$ upon the stochastic characteristics given from (3.3) to (3.9), which is directly given below without elaborating any further:

$$\hat{\mathbf{x}}(k) = \hat{\mathbf{x}}(k/k-1) + \mathbf{G}(k)\mathbf{d}(k/k-1) \tag{3.11}$$

$$\begin{aligned} \mathbf{D_{xx}}(k) = {} & [I - \mathbf{G}(k)\mathbf{C}(k)]\mathbf{D_{xx}}(k/k-1) \rightarrow \\ & [I - \mathbf{G}(k)\mathbf{C}(k)]^T + \mathbf{G}(k)\mathbf{R}(k)\mathbf{G}^T(k) \end{aligned} \tag{3.12}$$

with the *one-step predicted state vector*

$$\hat{\mathbf{x}}(k/k-1) = \mathbf{A}(k, k-1)\hat{\mathbf{x}}(k-1/k-1) \tag{3.13}$$

$$\mathbf{D_{xx}}(k/k-1)=\mathbf{A}(k,k-1)\mathbf{D_{xx}}(k-1/k-1)$$
$$\rightarrow \mathbf{A}^T(k,k-1)+\mathbf{B}(k,k-1)\mathbf{Q}(k)\mathbf{B}^T(k,k-1) \tag{3.14}$$

the *system innovation vector*

$$\mathbf{d}(k/k-1)=\mathbf{z}(k)-\mathbf{C}(k)\hat{\mathbf{x}}(k/k-1) \tag{3.15}$$
$$\mathbf{D}_{dd}(k/k-1)=\mathbf{C}(k)\mathbf{D_{xx}}(k-1/k-1)\mathbf{C}^T(k)+\mathbf{R}(k) \tag{3.16}$$

and the Kalman gain matrix

$$\mathbf{G}(k)=\mathbf{D_{xx}}(k/k-1)\mathbf{C}^T(k)\mathbf{D}_{dd}^{-1}(k/k-1) \tag{3.17}$$

In particular, the epochwise system innovation vector, the difference between the real and predicted measurement vectors as in (3.15), is of paramount importance to the error analysis in Kalman filtering. One can prove that the system innovation vectors in the series of $\mathbf{d}(1/0), \mathbf{d}(2/1), ..., \mathbf{d}(k/k-1), ...$ are independent of each other [35,36], namely, $Cov(\mathbf{d}(i/i-1), \mathbf{d}(j/j-1))=\mathbf{O}$ for $i \neq j$. Besides, the uncertainty of $\mathbf{d}(k/k-1)$ is obviously originating from the process noise series $\mathbf{w}(1), ..., \mathbf{w}(k), ...$, the measurement noise series $\Delta(1), ..., \Delta(k), ...$ along with the initial state vector. Usually, the error analysis in Kalman filtering has been centered on the system innovation series.

3.3 Alternate formulations of Kalman filter after least squares principle

As a matter of fact, the system and measurement models in (3.1) and (3.2) are associated with three groups of independent stochastic information that is propagated into the state solution from time to time. Specifically at epoch k, the system is contaminated by (i) the real measurement noise vector $\Delta(k)$, (ii) the system process noise vector $\mathbf{w}(k)$, and (iii) the noise associated with the predicted state vector $\hat{\mathbf{x}}(k/k-1)$ starting with $\mathbf{x}(0)$, into which $\Delta(1), ..., \Delta(k-1)$ and $\mathbf{w}(1), ..., \mathbf{w}(k-1)$ are propagated through the recursive mechanism as in (3.1) and (3.2).

Along two different paths, the Kalman filtering algorithm is equivalently derived following the least squares principle. First, under the consideration of (ii) and (iii) together, the predicted state vector is modeled as a pseudo-measurement vector as follows:

$$\bar{\mathbf{l}}_x(k)=\mathbf{A}(k,k-1)\hat{\mathbf{x}}(k-1)=\hat{\mathbf{x}}(k/k-1) \tag{3.18}$$

with its variance matrix

$$\mathbf{D}_{\bar{l}_x\bar{l}_x}(k) = \mathbf{A}(k, k-1)\mathbf{D}_{xx}(k-1)\mathbf{A}^T(k, k-1) + \mathbf{B}(k, k-1)\mathbf{Q}(k)\mathbf{B}^T(k, k-1) \tag{3.19}$$

so is the real measurement vector from (i):

$$\mathbf{l}_z(k) = \mathbf{z}(k) \quad \mathbf{R}(k) \tag{3.20}$$

(3.18) and (3.20) are further transited to their residual equations:

$$\mathbf{v}_{\bar{\mathbf{l}}_\mathbf{x}}(k) = \hat{\mathbf{x}}(k) - \bar{\mathbf{l}}_x(k) \qquad \mathbf{D}_{\bar{\mathbf{l}}_\mathbf{x}\bar{\mathbf{l}}_\mathbf{x}}(k) \tag{3.21}$$

$$\mathbf{v}_\mathbf{z}(k) = \mathbf{C}(k)\hat{\mathbf{x}}(k) - \mathbf{z}(k) \qquad \mathbf{R}(k) \tag{3.22}$$

In (3.21) and (3.22), there are n states and $(n+p)$ measurements. For a least squares solution, the goal function (as ***Alternate*** 1) runs accordingly:

$$g_{a_1}(k) = \mathbf{v}_{\bar{\mathbf{l}}_\mathbf{x}}^T(k)\mathbf{D}_{\bar{\mathbf{l}}_\mathbf{x}\bar{\mathbf{l}}_\mathbf{x}}^{-1}(k)\mathbf{v}_{\bar{\mathbf{l}}_\mathbf{x}}(k) + \mathbf{v}_z^T(k)\mathbf{R}^{-1}(k)\mathbf{v}_\mathbf{z}(k) = \min \tag{3.23}$$

which has widely and repeatedly been used to prove the equivalency between two principles (the minimum variance and the least squares) in deriving the Kalman filtering algorithm. In short, the least squares solution of the state vector after (3.23) is identical with the solution from (3.11) to (3.17). The repetitious details need not be given here.

Second, the pseudo-measurement vector in (3.18) is further split into two pseudo-measurement vectors:

$$\mathbf{l}_x(k) = \mathbf{A}(k, k-1)\hat{\mathbf{x}}(k-1) = \hat{\mathbf{x}}(k/k-1) \quad \mathbf{D}_{l_xl_x}(k) \tag{3.24}$$

$$\mathbf{l}_w(k) = \mathbf{w}_0(k) \qquad \mathbf{Q}(k) \tag{3.25}$$

with $\mathbf{w}_0(k) = \mathbf{o}$ normally and

$$\mathbf{D}_{\mathbf{l_x l_x}}(k) = \mathbf{A}(k, k-1)\mathbf{D}_{xx}(k-1)\mathbf{A}^T(k, k-1) \tag{3.26}$$

The real measurement vector remains as in (3.20).

In analogy to (3.21) and (3.22), their corresponding residual equations are given below:

$$\mathbf{v}_{\mathbf{l}_x}(k) = \hat{\mathbf{x}}(k) - \mathbf{B}(k, k-1)\hat{\mathbf{w}}(k) - \mathbf{l}_x(k) \tag{3.27}$$

$$\mathbf{v}_w(k) = \hat{\mathbf{w}}(k) - \mathbf{l}_w(k) \tag{3.28}$$

$$\mathbf{v_z}(k) = \mathbf{C}(k)\hat{\mathbf{x}}(k) \qquad\qquad - \mathbf{z}(k) \tag{3.29}$$

with $\mathbf{D}_{l_x l_x}(k)$, $\mathbf{Q}(k)$ and $\mathbf{R}(k)$ as their measurement variance matrices and $\mathbf{v}_{\mathbf{l}_w}(k) = \mathbf{v}_w(k)$, in which the state vector is extended to include the process noise vector $\mathbf{w}(k)$. In seeking a least squares solution for $\mathbf{x}(k)$ and $\mathbf{w}(k)$, a goal function (as ***Alternate*** 2) is constructed as follows:

$$\begin{aligned} g_{a_2}(k) &= \mathbf{v}_{\mathbf{l_x}}^T(k)\mathbf{D}_{\mathbf{l_x l_x}}^{-1}(k)\mathbf{v}_{\mathbf{l_x}}(k) \\ &\quad + \mathbf{v}_{\mathbf{l}_w}^T(k)\mathbf{Q}^{-1}(k)\mathbf{v}_{\mathbf{l}_w}(k) + \mathbf{v}_z^T(k)\mathbf{R}^{-1}(k)\mathbf{v_z}(k) = \min \end{aligned} \tag{3.30}$$

In (3.27), (3.28) and (3.22), there are $(n+m)$ states and $(n+m+p)$ measurements. The number of the redundant measurements remains unchanged, namely, p with either the ***Alternate*** 1 or the ***Alternate*** 2. It is not in question of the solution identity of the state vector $\mathbf{x}(k)$ after (3.30) and (3.23) [25]. The beauty with the ***Alternate*** 2 lies in the feasibility for the direct analysis of the original error sources inclusive of the residuals of the process noise vector together with the residuals of the real measurement vector and the computation of the redundancy contribution possessed by the process noise vector and the measurement vector. In case that $\mathbf{Q}(k)$ and/or $\mathbf{R}(k)$ are diagonal, one is able to calculate the individual redundancy indices for each factor in $\mathbf{w}(k)$ and/or for each measurement in $\mathbf{z}(k)$. Both of the residuals and the associated redundancy indices are of paramount importance to construct the test statistics and estimate variance components for refining the stochastic model.

3.4 Redundancy contribution in Kalman filtering

Baarda [1] initiated the reliability theory in the method of least squares. It consists of the internal reliability and the external reliability. The former is a measure of the system capability to detect measurement outliers with a specific probability while the latter is the model response to undetected model errors (systematic and measurement errors) [3]. Three measures are commonly used in internal reliability analysis [3]: (1) the redundancy contribution, which controls how a given error in a measurement affects its residual; (2) the minimal detectable outlier on a measurement at a significance level of α and with the test power of $1-\beta$; and (3) the maximum non-centrality parameter as a global measure, which is based on the quadratic form of the measurement residual vector. For further details about reliability analysis, refer to [1,3,17].

The reliability analytics was systematically introduced into Kalman filtering by [25]. Here, the discussion is limited to the redundancy contribution of measurements (inclusive of real and pseudo measurements) because it is distinctly the key of reliability analysis. Refer to [25,27] for more details on the subject.

Quantitatively, the redundancy contribution of a measurement vector is represented by the diagonal elements of the idempotent matrix $\mathbf{D}_{vv}\mathbf{D}_{ll}^{-1}$ originating from the well-known equation,

$$\mathbf{v_l} = \mathbf{D_{vv}}\mathbf{D_{ll}^{-1}}\mathbf{e_l} \tag{3.31}$$

where $\mathbf{e_l}$ and $\mathbf{v_l}$ are the error vector and estimated residual vector of a measurement vector $\mathbf{l}$, respectively, with the measurement variance matrix $\mathbf{D}_{ll}$ and the variance matrix $\mathbf{D}_{vv}$ of $\mathbf{v_l}$. In order to derive the redundancy contribution in Kalman filtering, the residual vectors given in (3.22), (3.27) and (3.28) are further expressed by the system innovation vector at epoch k:

$$\mathbf{v}_{\mathbf{l}_x}(k) = \mathbf{D}_{l_x l_x}(k)\mathbf{D}_{xx}^{-1}(k/k-1)\mathbf{G}(k)\mathbf{d}(k/k-1) \tag{3.32}$$

$$\mathbf{v}_w(k) = \mathbf{Q}(k)\mathbf{B}^T(k-1)\mathbf{D}_{xx}^{-1}(k/k-1)\mathbf{G}(k)\mathbf{d}(k/k-1) \tag{3.33}$$

$$\mathbf{v_z}(k) = [\mathbf{C}(k)\mathbf{G}(k) - \mathbf{I}]\mathbf{d}(k/k-1) \tag{3.34}$$

with their variance matrices

$$\begin{aligned}\mathbf{D}_{\mathbf{v}_{\mathbf{l}_x}\mathbf{v}_{\mathbf{l}_x}}(k) = {} & A(k,k-1)\mathbf{D}_{xx}(k-1)A^T(k,k-1)C^T(k) \rightarrow \\ & \mathbf{D}_{dd}^{-1}(k-1)C(k)A(k,k-1)\mathbf{D}_{xx}(k-1)A^T(k,k-1)\end{aligned} \tag{3.35}$$

$$\begin{aligned}\mathbf{D}_{\mathbf{v}_w\mathbf{v}_w}(k) = {} & Q(k)B^T(k,k-1)C^T(k) \rightarrow \\ & \mathbf{D}_{dd}^{-1}(k-1)C(k)B(k,k-1)Q(k)\end{aligned} \tag{3.36}$$

$$\mathbf{D}_{\mathbf{v}_z\mathbf{v}_z}(k) = [I - C(k)G(k)]R(k) \tag{3.37}$$

The redundancy contributions in measurement groups corresponding to (3.27), (3.28) and (3.22) are

$$\begin{aligned}r_{l_x}(k) = {} & trace\{A(k,k-1)\mathbf{D}_{xx}(k-1)A^T(k,k-1) \rightarrow \\ & C^T(k)\mathbf{D}_{dd}^{-1}(k-1)C(k)\}\end{aligned} \tag{3.38}$$

$$\begin{aligned}r_{l_w}(k) = {} & trace\{Q(k)B^T(k,k-1)C^T(k) \rightarrow \\ & \mathbf{D}_{dd}^{-1}(k-1)C(k)B(k,k-1)\}\end{aligned} \tag{3.39}$$

$$r_z(k) = trace\{I - C(k)G(k)\} \tag{3.40}$$

For the entire system either after (3.1) and (3.2), after (3.21) and (3.22), or after (3.27), (3.28) and (3.22), the total redundancy number at epoch k satisfies [25,27,4]

$$r(k) = r_{l_x}(k) + r_{l_w}(k) + r_z(k) = p(k) \tag{3.41}$$

wherein $p(k)$ is the number of the real measurements or the dimension of $\mathbf{z}(k)$.

In practice, $\mathbf{Q}(k)$ and $\mathbf{R}(k)$ are commonly diagonal so that the individual redundancy indices in components for the process noise vector are

$$r_{w_i}(k) = \{Q(k)B^T(k, k-1)C^T(k)\mathbf{D}_{dd}^{-1}(k-1) \rightarrow C(k)B(k, k-1)\}_{ii}(i = 1, 2, ..., m(k)) \tag{3.42}$$

and for the measurement vector

$$r_{z_i}(k) = \{I - C(k)G(k)\}_{ii}(i = 1, 2, ..., p(k)) \tag{3.43}$$

Indeed, $\mathbf{D}_{\mathbf{l_x l_x}}(k)$ in (3.26) is not a diagonal matrix in general, no individual redundancy indices in components are here defined for $\mathbf{l}_x(k)$. For more about the reliability measures for the correlated measurements one refers to [17,37].

3.5 Variance of unit weight and variance component estimation

3.5.1 Variance of unit weight and posteriori variance matrix of $\hat{x}(k)$

The variance of unit weight denoted by σ_0^2, also as reference variance or variance factor, is well known in surveying and mapping, where it is mostly estimated using the measurement residuals in order to posteriorly assess the accuracy of the least squares solution. However, it is not true in the community in which a Kalman filter is applied at large, where a great confusion or mismatch is widely spread. The variance matrix $\mathbf{D_{xx}}(k)$ in (3.12) is characterized as the posterior variance. Indeed, it is simply carried out recursively through the variance propagation and never involves either any measurement residuals or system innovations. At epoch k, the posterior variance matrix of the estimated state vector at epoch k is

$$\hat{\mathbf{D}}_{\mathbf{xx}}(k) = \hat{\sigma}_0^2(k)\mathbf{D}_{\mathbf{xx}}(k) \tag{3.44}$$

wherein $\hat{\sigma}_0^2(k)$ is the local variance of unit weight epochwise and is estimated by

$$\hat{\sigma}_0^2(k) = \frac{\mathbf{d}^T(k/k-1)\mathbf{D}_{dd}^{-1}(k/k-1)\mathbf{d}(k/k-1)}{p(k)} \tag{3.45a}$$

after the model in Section 3.2, or

$$\hat{\sigma}_0^2(k) = \frac{\mathbf{v}_{\bar{\mathbf{l}}_\mathbf{x}}^T(k)\mathbf{D}_{\bar{\mathbf{l}}_\mathbf{x}\bar{\mathbf{l}}_\mathbf{x}}^{-1}(k)\mathbf{v}_{\bar{\mathbf{l}}_\mathbf{x}}(k) + \mathbf{v}_\mathbf{z}^T(k)\mathbf{R}^{-1}(k)\mathbf{v}_\mathbf{z}(k)}{p(k)} \tag{3.45b}$$

after the Alternate 1 from (3.21) and (3.22) in Section 3.3, or

$$\hat{\sigma}_0^2(k) = \frac{\mathbf{v}_{\mathbf{l}_\mathbf{x}}^T(k)\mathbf{D}_{\mathbf{l}_\mathbf{x}\mathbf{l}_\mathbf{x}}^{-1}(k)\mathbf{v}_{\mathbf{l}_\mathbf{x}}(k) + \mathbf{v}_{\mathbf{l}_w}^T(k)\mathbf{Q}^{-1}(k)\mathbf{v}_{\mathbf{l}_w}(k) + \mathbf{v}_\mathbf{z}^T(k)\mathbf{R}^{-1}(k)\mathbf{v}_\mathbf{z}(k)}{p(k)} \tag{3.45c}$$

after the Alternate 2 from (3.27), (3.28) and (3.22). The equivalence between (3.45a) and (3.45b) was proved by [38,39] while the equivalence between (3.45a) and (3.45c) was given in [25].

By taking the advantage of the independency of the system innovation vectors or the measurement residual vectors from epoch to epoch, σ_0^2 may be also estimated over a specific time interval as its regional estimate, e.g. corresponding to (3.45a), one has

$$\hat{\sigma}_{0(regional)}^2 = \frac{\sum_{i=k-s+1}^{k} \mathbf{d}^T(i/i-1)\mathbf{D}_{dd}^{-1}(i/i-1)\mathbf{d}(i/i-1)}{\sum_{i=k-s+1}^{k} p(i)} \tag{3.46}$$

Over $(k-s+1, \ldots, k)$ or even over the whole time duration associated with a dataset as its global estimate,

$$\hat{\sigma}_{0(global)}^2 = \frac{\sum_{i=1}^{k} \mathbf{d}^T(i/i-1)\mathbf{D}_{dd}^{-1}(i/i-1)\mathbf{d}(i/i-1)}{\sum_{i=1}^{k} p(i)} \tag{3.47}$$

over $(1, 2, \ldots, k)$. The choice selection depends on user's requirement either in order to reflect the variation of the reference variances locally, or to estimate it relatively stable regionally, or overall globally [25,26].

Incidentally, Koch [12] modeled σ_0^2 based on the normal-Gamma distribution as one of the states in Kalman filtering.

3.5.2 Estimation of variance components

The estimation of variance components in Kalman filtering has attracted considerable attention since 1970s. Intensive literature reviews on this subject can be found in [25,27,13]. In general, there exist two classes of methods: variance component matrix estimation (VCME) and variance component (scalar) estimation (VCE).

As a pioneer, Mehra [20,21] presented his VCME work for estimating **Q** and ***R*** (defined in (3.3) and (3.4)) using the system innovation series in steady-state Kalman filtering in four different ways: Bayesian, maximum likelihood, correlation and covariance matching. His work is still being adopted in a number of the works nowadays ([6,2,19] etc.).

The various versions of Helmert's VCE method [10] developed in least squares have been more attractive and popular in comparison. For more details on the VCE methods in least squares, refer to [40–45,3,33,22,5]. Nowadays, they have widely been applied in GNSS and other applications ([46–51,56,9,7] etc.).

Meanwhile, different VCE methods have been developed in the process of Kalman filtering ([32,25,4,11,31,29,13,14,34] etc.). Specifically, most of these VCE algorithms after Helmert in Kalman filtering were based on the formulation of (3.21) and (3.22). Among them, in particular Yu et al. [32] developed the MINQE algorithm to estimate three variance factors corresponding to the predicted state vector, the process noise vector and the real measurement vector, and applied it in monitoring leveling networks with seven epochs of measurements. Furthermore, upon (3.22), (3.27), and (3.30), Wang [25] developed a VCE method not only for the same three variance factors and also for the individual variance components in $\mathbf{Q}(k)$ and $\mathbf{R}(k)$ in the case they are diagonal. Especially, by taking the advantage of the individual redundant indices as given in (3.42) and (3.43), the simplified VCE algorithm after [41,18] was transplanted into Kalman filtering by [25, 4] (either epochwise or with multiple epochs) for ($k_1 \leq k \leq k_2$):

$$\sigma^2_{w_i(k)} = \frac{v^2_{l_{w_i}}(k)}{r_{w_i(k)}} \quad \text{or} \quad \sigma^2_{w_i(k_1,\dots,k_2)} = \frac{\sum_{k=k_1}^{k_2} v^2_{l_{w_i}}(k)}{\sum_{k=k_1}^{k_2} r_{l_{w_i}(k)}} \tag{3.48}$$

for each element in the process noise vector $\mathbf{w}(k)$ and

$$\sigma^2_{z_i(k)} = \frac{v^2_{z_i}(k)}{r_{z_i(k)}} \quad \text{or} \quad \sigma^2_{z_i(k_1,\dots,k_2)} = \frac{\sum_{k=k_1}^{k_2} v^2_{z_i}(k)}{\sum_{k=k_1}^{k_2} r_{z_i(k)}} \tag{3.49}$$

for each measurement in the measurement vector $\mathbf{z}(k)$. The estimated variances after (3.48) and (3.49) approximate their rigorous Helmert estimators well in the applications where a large number of observation epochs are available, e.g. kinematic GNSS positioning and/or multisensor integrated navigation [25,28] similar to the case in Photogrammetry [41,18,17]. Due to space limit, no details are further provided here.

3.6 Test statistics

A Kalman filter functions properly only if the assumption about its model structures, process and measurement noise characteristics are correct or realistic (refer to Section 3.2). An improper model, false modeled process noise, false modeled measurement noise or unexpected sudden changes of the state vector may result in divergent filtering solution. Statistic hypothesis testing belongs to the paramount aspects of error analysis in Kalman filtering. Traditionally, the statistic algorithms in Kalman filtering can be divided into two categories: multiple hypothesis filter detectors [52–54] and innovation-based detection ([55,35,23] etc.). Based the formulation of (3.22), (3.27) and (3.28), the test statistics were constructed not only on the basis of normal and χ^2-distributions, but also on the basis of t- and F-distributions using both of the system innovation and measurement residuals in Kalman filter [25,26], which are summarized in this subsection.

All of the following test statistics are constructed under the stochastic assumptions as in Section 3.2. At an arbitrary epoch k, one can straightforwardly derive the following (after the law of error propagation):

$$\mathbf{d}(k/k-1) \sim N(o, \mathbf{D}_{dd}(k/k-1)) \quad ((3.15) \text{ and } (3.16)) \tag{3.50}$$

$$\mathbf{v}_w(k) \sim N(o, \mathbf{D}_{\mathbf{v}_w\mathbf{v}_w}(k)) \quad ((3.33) \text{ and } (3.36)) \tag{3.51}$$

$$\mathbf{v}_z(k) \sim N(o, \mathbf{D}_{\mathbf{v}_z\mathbf{v}_z}(k)) \quad ((3.34) \text{ and } (3.37)) \tag{3.52}$$

$$\begin{aligned}&\mathbf{d}^T(k/k-1)\mathbf{D}_{dd}^{-1}(k/k-1)\mathbf{d}(k/k-1)\\&\quad = \mathbf{v}_{\mathbf{l_x}}^T(k)\mathbf{D}_{\mathbf{l_x l_x}}^{-1}(k)\mathbf{v}_{\mathbf{l_x}}(k) + \mathbf{v}_{\mathbf{l}_w}^T(k)\mathbf{Q}^{-1}(k)\mathbf{v}_{\mathbf{l}_w}(k)\\&\qquad + \mathbf{v}_{\mathbf{z}}^T(k)\mathbf{R}^{-1}(k)\mathbf{v}_{\mathbf{z}}(k) \sim \chi^2(p(k))\end{aligned} \tag{3.53}$$

Further, the individual components in $\mathbf{d}(k/k-1)$, $\mathbf{v}_w(k)$ and $\mathbf{v}_z(k)$ conform to the specific one-dimensional normal distributions and their own series over a certain specified time interval from epoch $(k-s+1)$ to epoch k conform to the s-dimensional normal distributions, too, due to $Cov(\mathbf{d}(i/i-1), \mathbf{d}(j/j-1)) = \mathbf{O}$ and $Cov(\mathbf{v}(i), \mathbf{v}(j)) = \mathbf{O}$ for $i \neq j$.

Hence, one built up three levels of test statistics for the system innovations and process noise and measurement residuals as follows [25,26].

Global test statistics are introduced right after the first k epochs are processed i) with all of the system innovation information for their individual components and for themselves by constructing the corresponding χ^2- test statistics ($k = 1, 2, ..., N$):

$$\sum_{j=1}^{k} \frac{\mathbf{d}_i^2(j/j-1)}{\sigma^2_{\mathbf{d}_i(j/j-1)}} \sim \chi^2(k) \tag{3.54}$$

$$\sum_{j=1}^{k} \mathbf{d}^T(j/j-1)\mathbf{D}_{dd}^{-1}(j/j-1)\mathbf{d}(j/j-1) \sim \chi^2(\sum_{j=1}^{k} p(j)) \tag{3.55}$$

and ii) with all of the process noise and measurement residuals for their individual components and for themselves by constructing the corresponding χ^2-test statistics ($k = 1, 2, ..., N$):

$$\sum_{j=1}^{k} \frac{v_{u_i}^2(j)}{\sigma^2_{v_{u_i}(j)}} \sim \chi^2(k)(\mathbf{u} = \mathbf{w}, \mathbf{z}) \tag{3.56}$$

$$\sum_{j=1}^{k} \mathbf{v}_u^T(j)\mathbf{D}_{v_u v_u}^{-1}(j)\mathbf{v}_u(j) \sim \chi^2(\sum_{j=1}^{k} r_u(j)) \tag{3.57}$$

where $r_u(j)$ is the total redundancy indices of $\mathbf{w}(j)$ in (3.39) or $\mathbf{z}(j)$ in (3.40).

Regional test statistics are similarly introduced for a certain specified time interval from epoch $(k - s + 1)$ to epoch k right after the first k epochs are processed i) with all of the system innovation information for their individual components and for themselves by constructing the corresponding χ^2- test statistics ($k = s, s + 1, ..., N$):

$$\sum_{j=k-s+1}^{k} \frac{\mathbf{d}_i^2(j/j-1)}{\sigma^2_{\mathbf{d}_i(j/j-1)}} \sim \chi^2(s) \tag{3.58}$$

$$\sum_{j=k-s+1}^{k} \mathbf{d}^T(j/j-1)\mathbf{D}_{dd}^{-1}(j/j-1)\mathbf{d}(j/j-1) \sim \chi^2(\sum_{j=k-s+1}^{k} p(j)) \tag{3.59}$$

and ii) with all of the process noise and measurement residuals for their individual components and for themselves by constructing the corresponding χ^2- test statistics ($k = s, s + 1, ..., N$):

$$\sum_{j=k-s+1}^{k} \frac{v_{u_i}^2(j)}{\sigma^2_{v_{u_i}(j)}} \sim \chi^2(s)(\mathbf{u} = \mathbf{w}, \mathbf{z}) \tag{3.60}$$

$$\sum_{j=k-s+1}^{k} \mathbf{v}_u^T(j)\mathbf{D}_{v_u v_u}^{-1}(j)\mathbf{v}_u(j) \sim \chi^2\left(\sum_{j=k-s+1}^{k} r_u(j)\right) \tag{3.61}$$

Here, additional F-test statistics are possibly further constructed as the ratio of the weighted residuals (or system innovations) squared from the first $(k-s)$ epochs to the weighted residuals (or system innovations) squared from the latest s epochs. For more details, refer to [25,26].

Local test statistics are constructed epochwise at an arbitrary epoch k either with the system innovation vector for their individual components and for themselves by constructing the corresponding χ^2-test statistics ($k = 1, 2, \ldots, N$):

$$\frac{\mathbf{d}_i(k/k-1)}{\sigma_{\mathbf{d}_i(k/k-1)}} \sim N(0,1) \tag{3.62}$$

$$\frac{\mathbf{d}_i(k/k-1)/\sigma_{\mathbf{d}_i(k/k-1)}}{\hat{\sigma}^2_{0(regional)}(k-s, \ldots, k-1)} \sim t(r_{regional}) \tag{3.63}$$

$$\mathbf{d}^T(k/k-1)\mathbf{D}_{dd}^{-1}(k/k-1)\mathbf{d}(k/k-1) \sim \chi^2(p(k)) \tag{3.64}$$

$$\frac{\mathbf{d}^T(k/k-1)\mathbf{D}_{dd}^{-1}(k/k-1)\mathbf{d}(k/k-1)/p(k)}{\hat{\sigma}^2_{0(regional)}(k-s, \ldots, k-1)} \sim F(p(k), r_{regional}) \tag{3.65}$$

where $\hat{\sigma}^2_{0(regional)}$ is the estimated variance of unit weight over a specific interval ($s-1$ epochs) before epoch k and $r_{regional}$ is the corresponding degrees of freedom. In (3.65), one can also select the whole past time interval up to $k-1$. The same test statistics are applicable to the residual vectors of the process noise vector and measurement vector for their individual components and for themselves ($k = 1, 2, \ldots, N$):

$$\frac{v_{u_i}(k)}{\sigma_{v_{u_i}(k)}} \sim N(0,1) \tag{3.66}$$

$$\frac{v_{u_i}(k)/\sigma_{v_{u_i}(k)}}{\hat{\sigma}^2_{0(regional)}(k-s, \ldots, k-1)} \sim t(r_{regional}) \tag{3.67}$$

$$\mathbf{v}_u^T(k)\mathbf{D}_{v_u v_u}^{-1}(k)\mathbf{v}_u(k) \sim \chi^2(r_u(k)) \tag{3.68}$$

$$\frac{\mathbf{v}_u^T(k)\mathbf{D}_{v_u v_u}^{-1}(k)\mathbf{v}_u(k)/r_u(k)}{\hat{\sigma}^2_{0(regional)}(k-s, \ldots, k-1)} \sim F(r_u(k), r_{regional}) \tag{3.69}$$

where $u = w, z$, $r_u(k)$ is the total redundancy indices of $\mathbf{w}(k)$ in (3.39) or $\mathbf{z}(k)$ in (3.40).

3.7 Real data analysis with multi-sensor integrated kinematic positioning and navigation

3.7.1 Overview

The dataset analyzed here came from a kinematic trajectory acquired by York University Multisensor Integrated System (YUMIS) on July 27, 2014 [16,14]. The data fusion is based on the generic multisensor integration strategy developed at York University ([14,15,30] etc.). Due to the space limit, only the results from an 11 minutes long stretch within this trajectory are presented.

Specifically, the state vector consists of the 3D position, velocity and acceleration, attitude angle, angular rate, accelerometer scale factor and bias, and gyro scale factor and bias vectors (27 states in total), while the process nose vector consists of 18 factors: the jerk vector, 3D changes of angular rates, accelerometer bias and scale factor and gyro bias and scale factor vectors. The used measurements are the double-differenced GPS L1 C/A and L1 carrier phases and the IMU specific force and angular rate vectors. The body frame was defined as forward-eastward-down (xyz).

3.7.2 Results

The results of the Kalman filter are summarized in 17 figures.

First, Fig. 3.1 to Fig. 3.4 overview the stretch of the trajectory presented here. The 2D top view and the vertical profile of the trajectory stretch are presented in Fig. 3.1 while the absolute velocity profile and the absolute acceleration profile are plotted in Fig. 3.2 and Fig. 3.3, respectively. Fig. 3.4 shows the profile of the attitude angles (roll, pitch and heading).

Second, the posterior standard deviations, system innovations and residuals (measurements and process noises) are shown from Fig. 3.5 to Fig. 3.14. The posteriori standard deviations of the estimated positions in north, east and down are plotted in Fig. 3.5 while the posteriori standard deviations of the estimated roll, pitch and heading angles are plotted in Fig. 3.6. Fig. 3.7 and Fig. 3.8 present the system innovation and residuals of the GPS L1 C/A and carrier phase measurements, respectively. The system innovation and residuals of the IMU specific forces and angular rates are given in Fig. 3.9 and Fig. 3.10, respectively. The residuals of the process noises (jerks) and the residuals of the change of the angular rates (as process noises) are shown in Fig. 3.11 and Fig. 3.12, respectively. Fig. 3.13 shows the residuals of the changes of IMU accelerometer and gyro biases (process noise) while

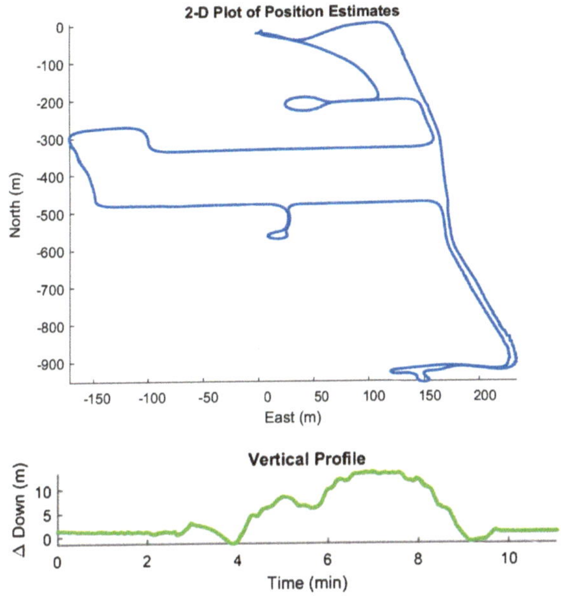

Figure 3.1 The 2D top view (top) and vertical profile (bottom) of the trajectory.

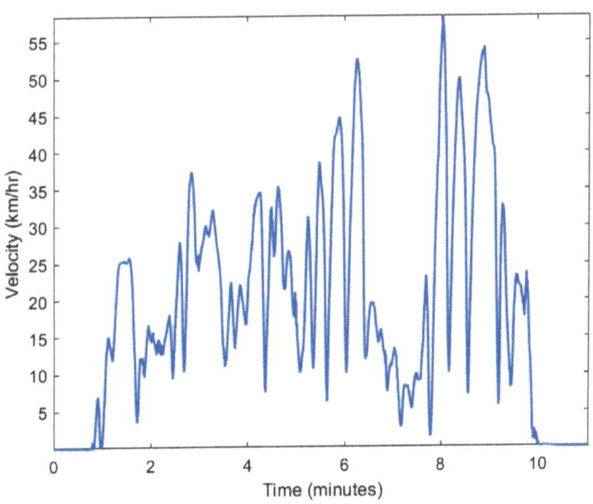

Figure 3.2 The profile of the absolute velocity.

the residuals of the changes of IMU accelerometer and gyro scale factors (process noise) are given in Fig. 3.14.

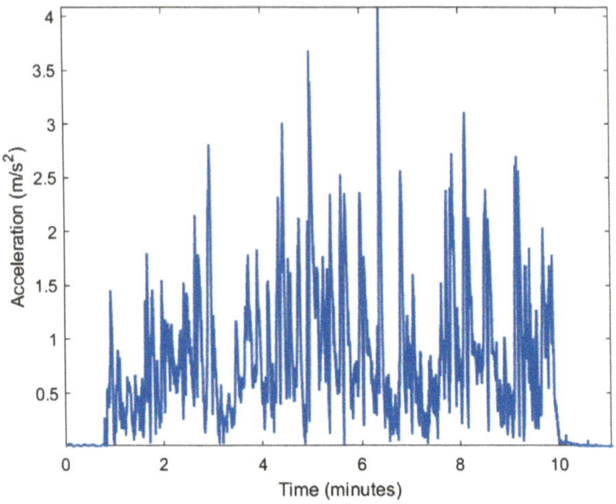

Figure 3.3 The profile of the absolute acceleration.

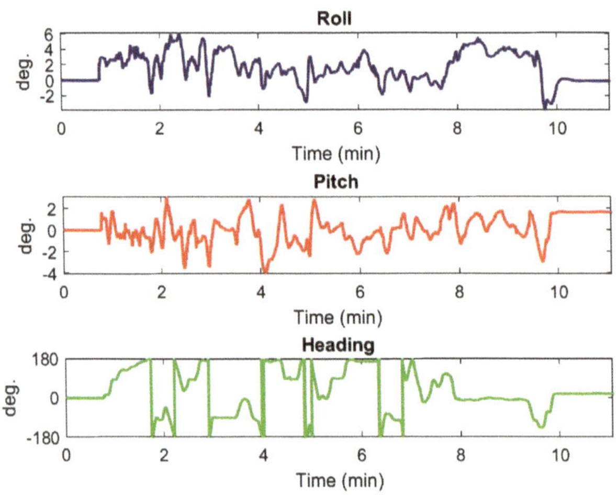

Figure 3.4 The profile of the attitude: roll-pitch-heading.

Third, Figs. 3.15, 3.16 and 3.17 give a good snapshot of the VCE purposely conducted as part of the comprehensive error analysis in Kalman filter. The results of the estimated variance components for GPS L1 C/A, GPS L1Carrier Phases, IMU specific forces and IMU angular rates are plotted in Fig. 3.15 while Fig. 3.16 gives six plots corresponding to the results of the estimated variance components for the process noises (jerks, changes of accelerometer biases, changes of angular rates, changes of gyro biases,

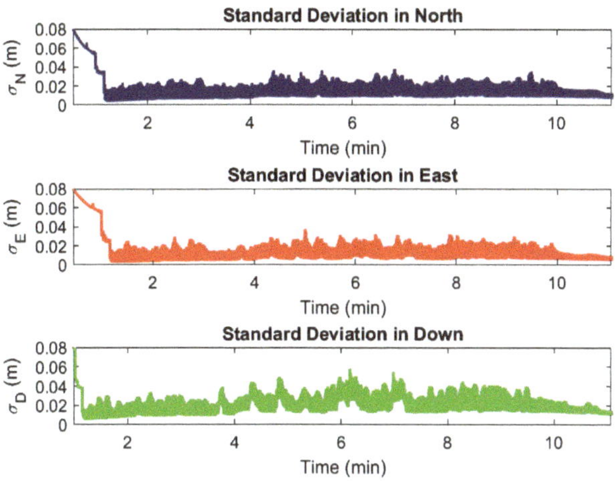

Figure 3.5 The quality of the position solution.

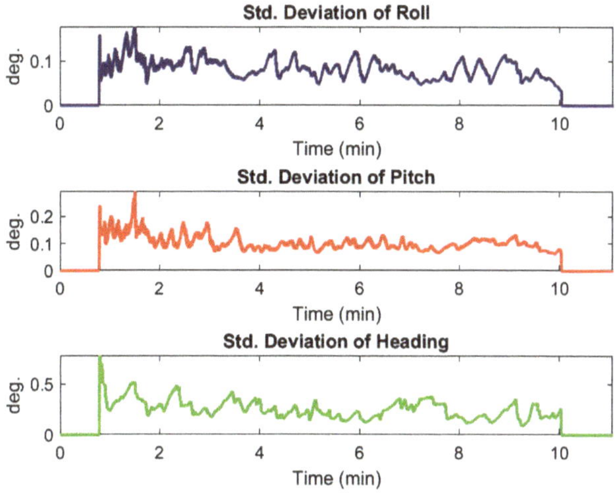

Figure 3.6 The quality of the attitude angles.

changes of accelerometer scale errors and changes of gyro scale errors in sequence). The redundancy indexes (the total, the subtotal of the real measurements, the subtotal of predicted states, the subtotal of process noises) are shown in four plots as in Fig. 3.17.

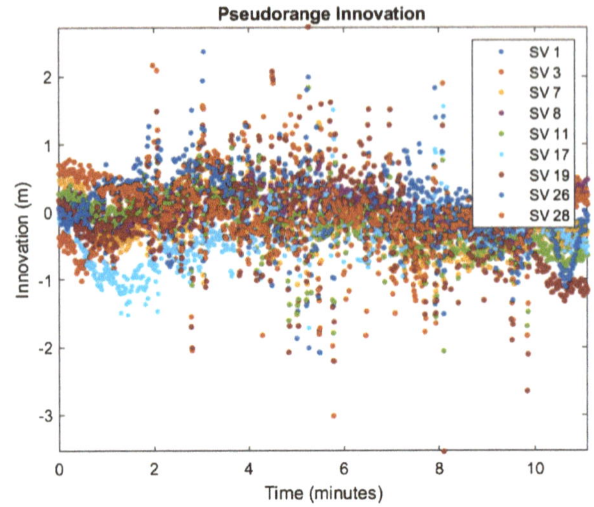

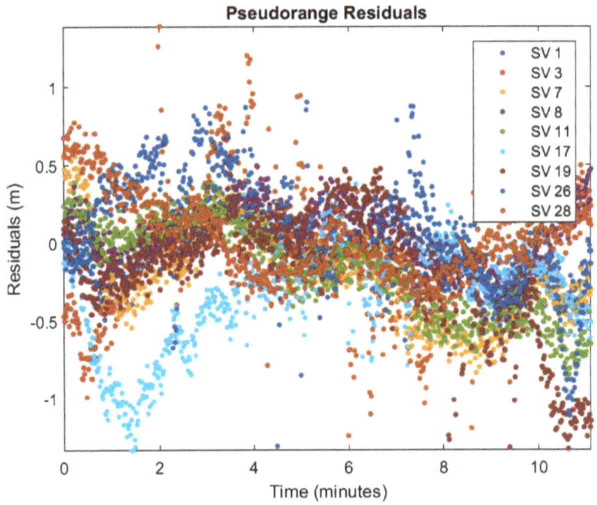

Figure 3.7 The GPS L1C/A innovations and residuals.

3.8 Remarks

This manuscript overviews the generic essentials to comprehensive error analysis in Kalman filter. Specifically, it is carried out in four aspects: (a) how to observe the error sources and execute their analysis (Section 3.3); (b) the system redundancy contribution for the predicted state vector, the process noise vector and the measurement vector and the individual redundancy

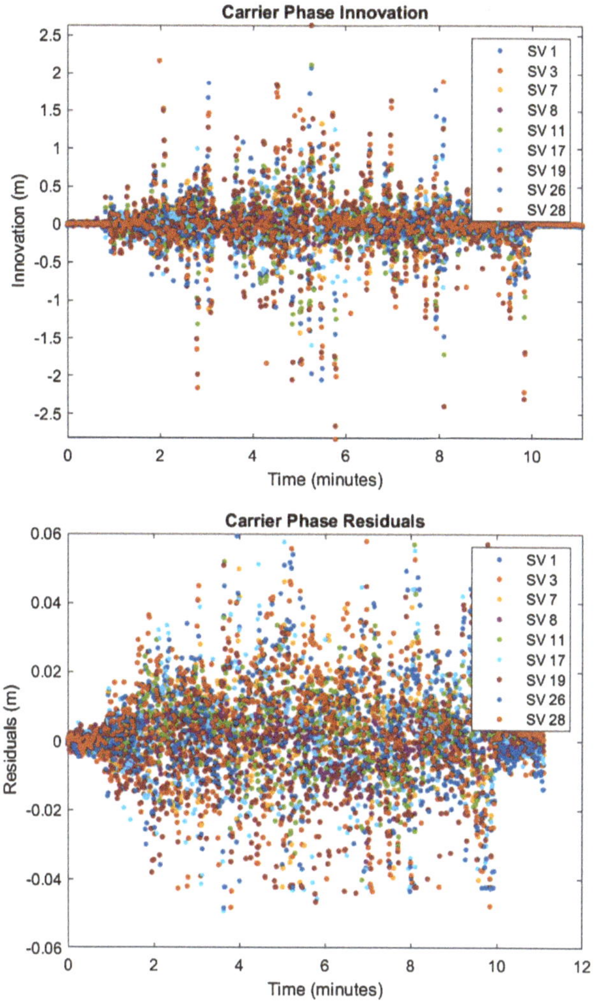

Figure 3.8 The GPS L1 carrier phase innovations and residuals.

indices for the independent individual process noise factors and the independent individual measurements (Section 3.4); (c) the variance of unit weight and variance component estimation (Section 3.5); and (d) the test statistics for the system innovations, the residuals of the process noise vectors and the measurement vectors (Section 3.6). Accordingly, an application is provided using a real dataset from the multisensor integrated navigation in Section 3.7. The authors find them to be of paramount importance to high precision and high accuracy applications in kinematic positioning

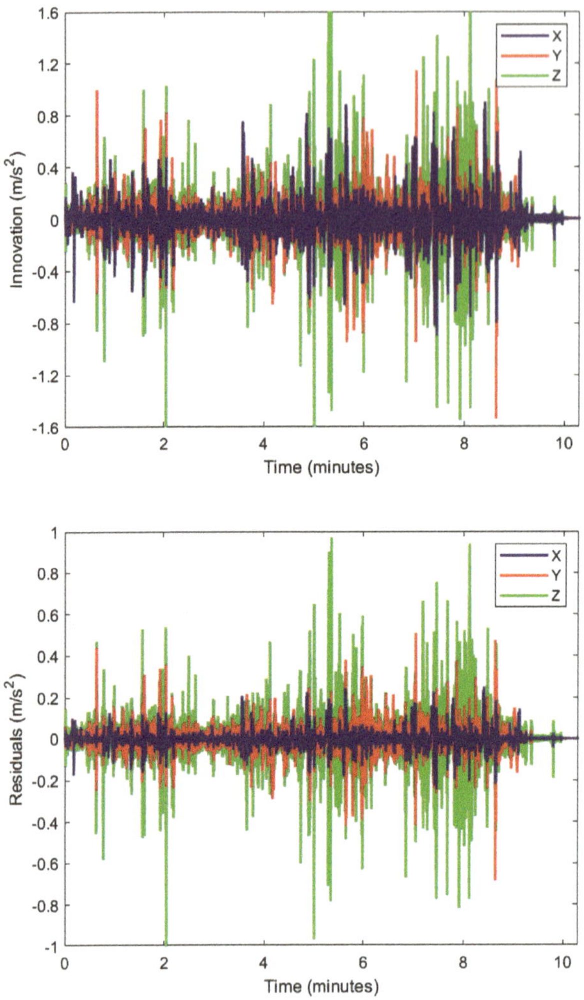

Figure 3.9 The innovation and residuals of IMU specific forces.

and navigation when it comes to centimeters level of positional accuracy requirements either in direct-georeferencing, the key to the modern automatic geospatial data acquisition, to the autonomous driving, or even to the modern multisensor integrated navigation. However, this has not comprehensively been conducted in academic and high-tech research and development yet. It is our expectation that such comprehensive error analysis in Kalman filtering becomes a standard process as in our least squares methods.

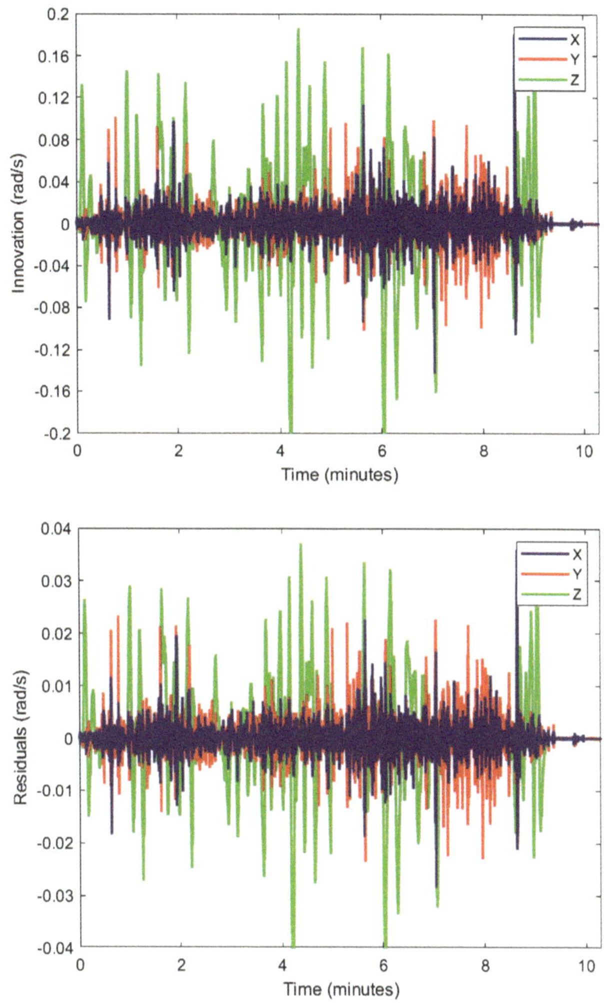

Figure 3.10 The innovation and residuals of IMU angular rates.

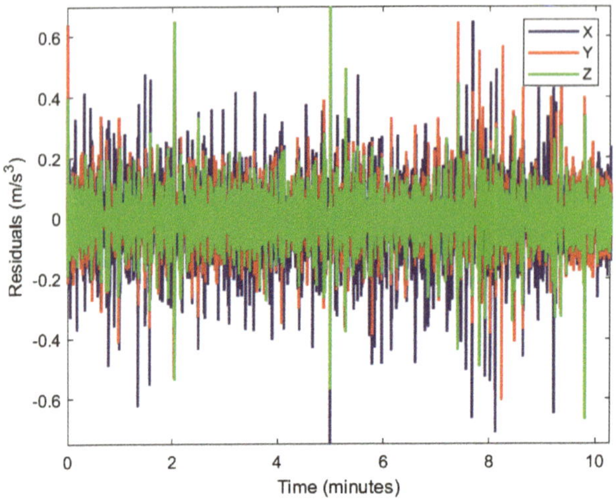

Figure 3.11 The residuals of jerks (process noise).

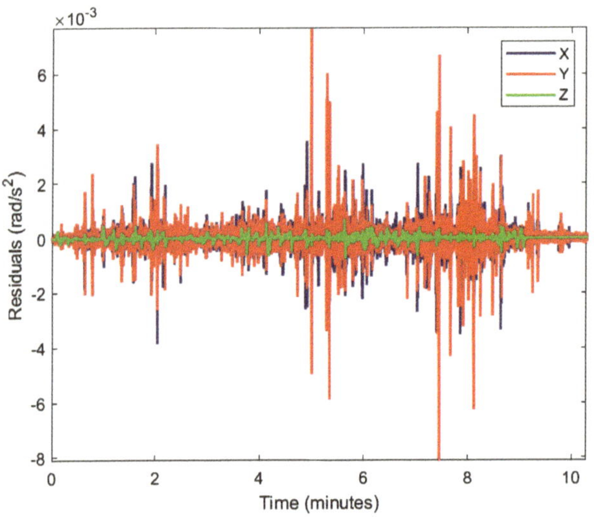

Figure 3.12 The residuals of the changes of the angular rates (process noise).

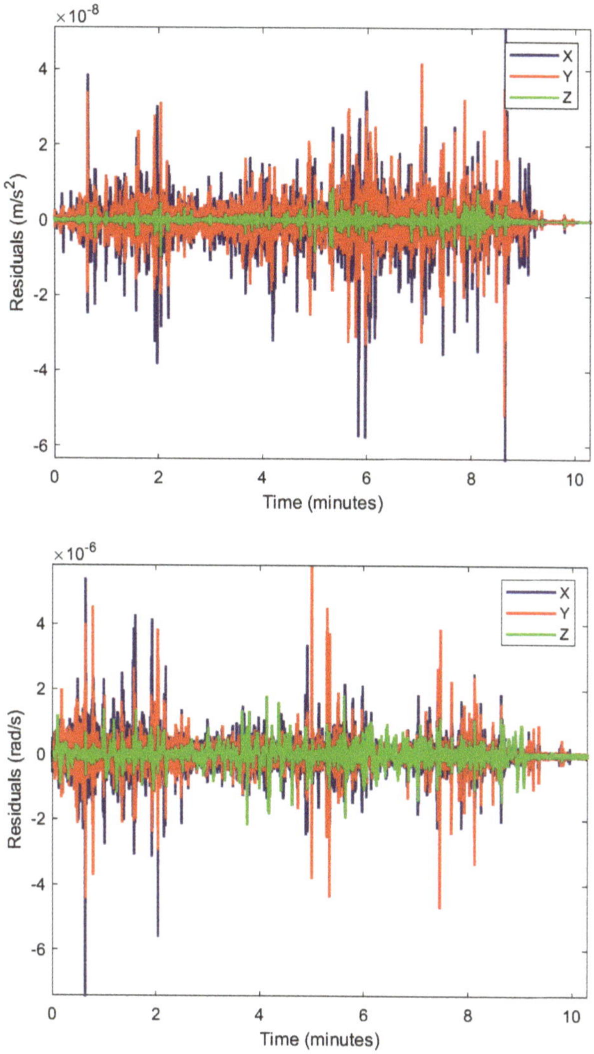

Figure 3.13 The residuals of the changes of IMU accelerometer and gyro biases (process noise).

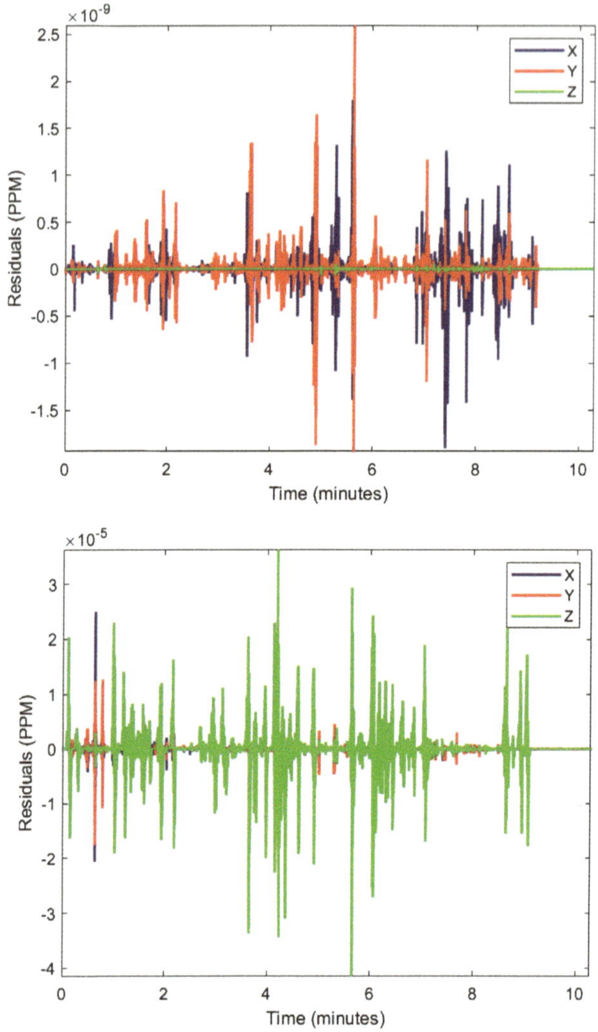

Figure 3.14 The residuals of the changes of IMU accelerometer and gyro scale factors (process noise).

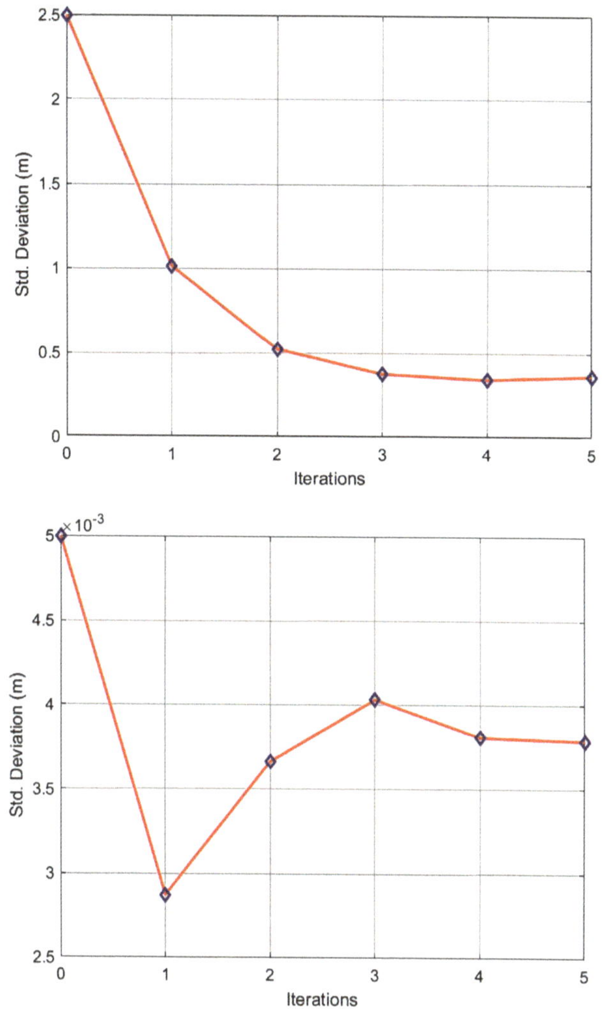

Figure 3.15 The variance component estimation of measurements.

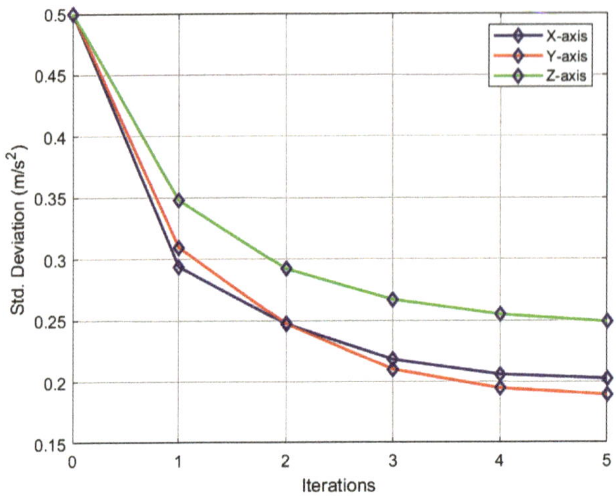

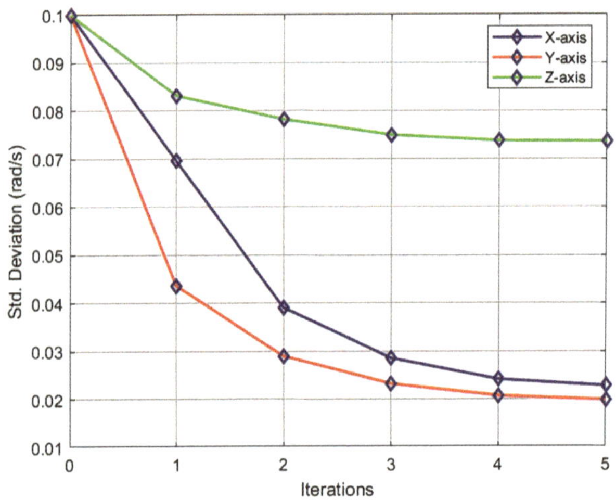

Figure 3.15 (*continued*)

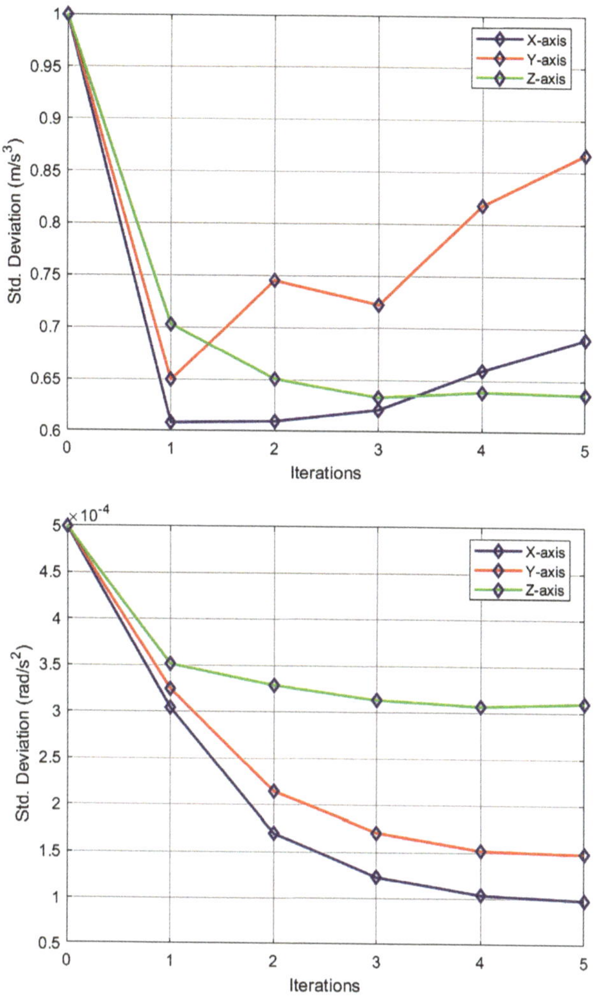

Figure 3.16 The variance component estimation of process noise factors.

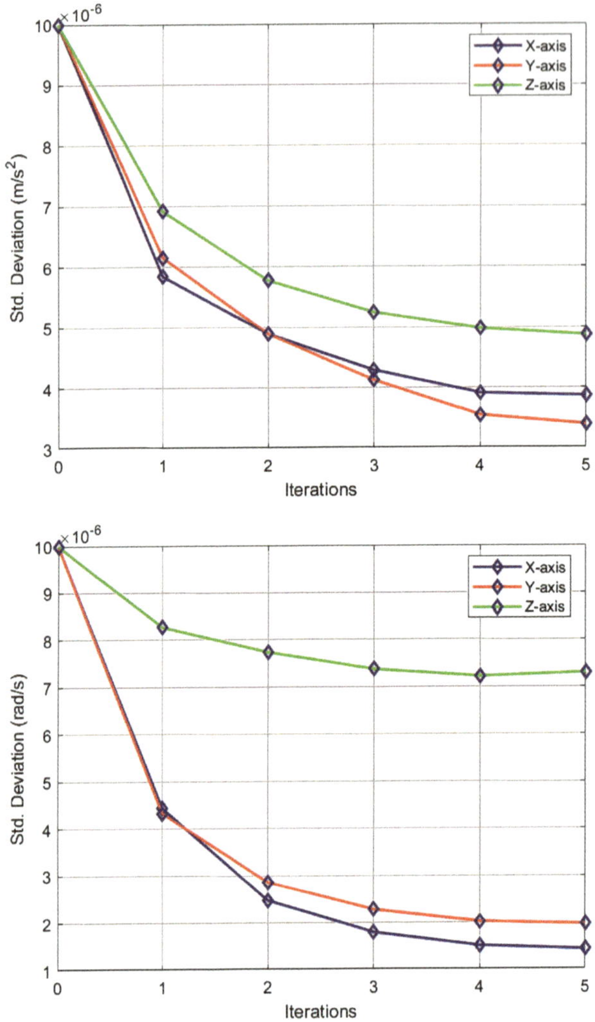

Figure 3.16 *(continued)*

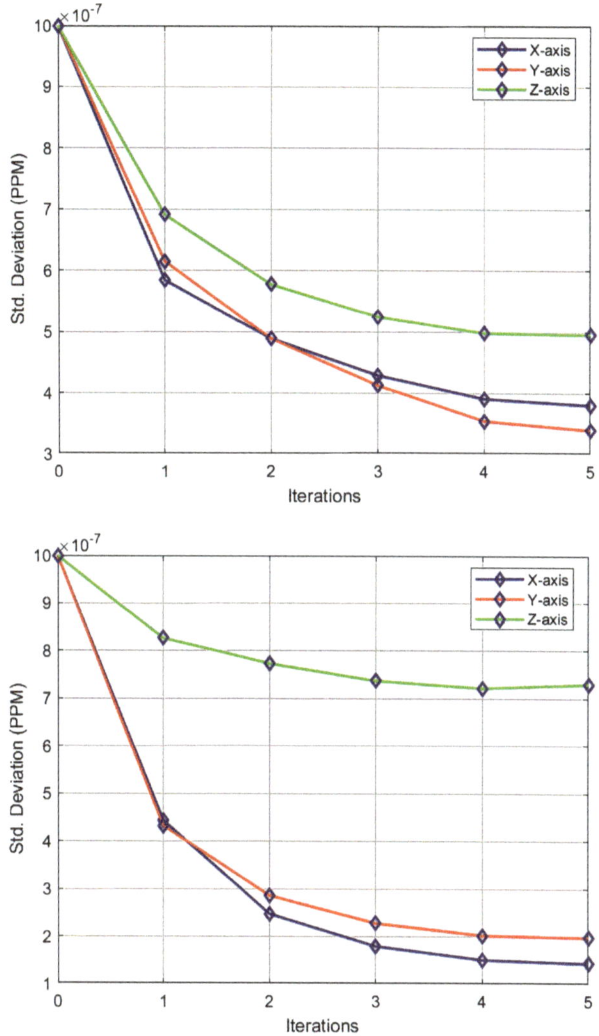

Figure 3.16 (*continued*)

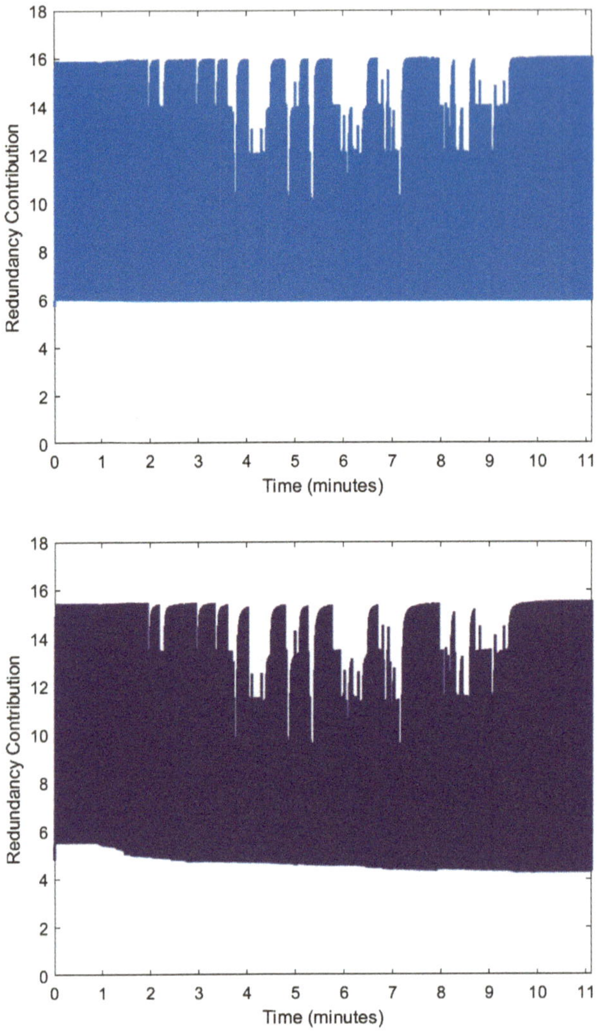

Figure 3.17 The redundancy indices (the total, real measurements, predicted states, process noises).

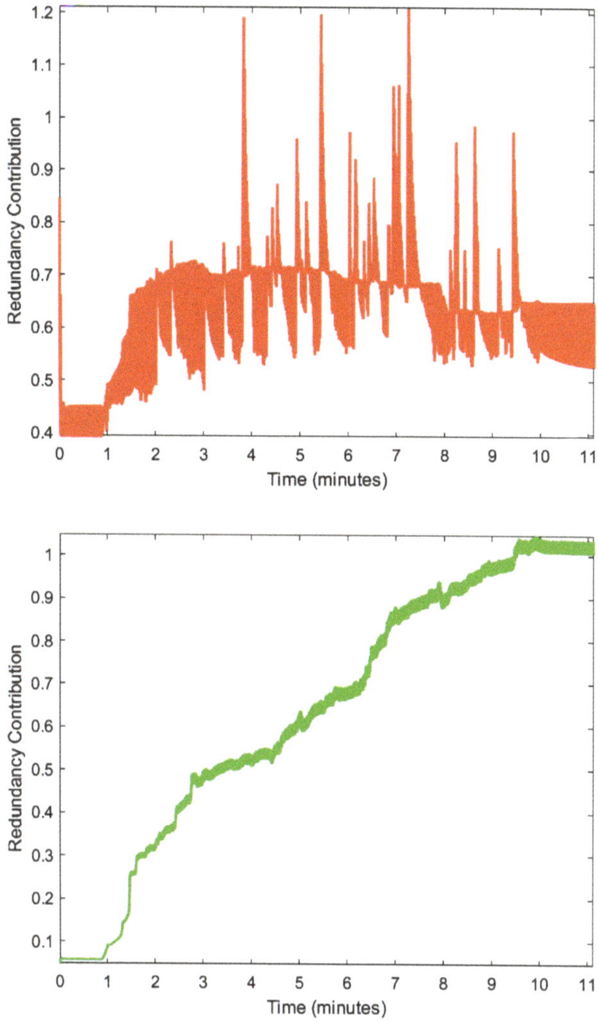

Figure 3.17 (*continued*)

References

[1] Willem Baarda, A Testing Procedure for Use in Geodetic Networks, Publications on Geodesy, No. 5, Vol. 2, NGC, Delft, Netherlands, 1968.

[2] V.A. Bavdekar, A.P. Deshpande, S.C. Patwardhan, Identification of process and measurement noise covariance for state and parameter estimation using extended Kalman filter, Journal of Process Control 21 (4) (2011) 585–601.

[3] Wilhelm Caspary, Concepts of Network and Deformation Analysis, Monograph 11, School of Surveying, University of New South Wales, Kensington, N.S.W., Australia, 1988.

[4] Wilhelm Caspary, Jianguo Wang, Redundanzanteile und Varianzkomponentenim Kalman Filter, Zeitschrift fuer Vermessusgswesen (ZFV) 123 (4) (1998) 121–128.
[5] X. Cui, Z. Yu, B. Tao, Generalized Surveying Adjustment (广义测量平差), WTUSM (Wuhan Technical University of Surveying and Mapping) Press, 2001.
[6] J. Dunık, M. Šimandl, Estimation of state and measurement noise covariance matrices by multi-step prediction, in: Proceedings of the 17th IFAC World Congress, 2008, pp. 3689–3694.
[7] Xiao Gao, Wujiao Dai, 抗差 Helmert 方差分量估计在 GPS/BDS 组合定位中的应用, 大地测量方差分与地球动力学, 34(1) (2014) 173–176.
[8] Arthur Gelb, et al., Applied Optimal Estimation, The Analytic Sciences Corporation & The M.I.T. Press, Cambridge Massachusetts London, 1974.
[9] Nilesh Gopaul, Jianguo Wang, Jiming Guo, Improving of GPS observation quality assessment through posteriori variance–covariance component estimation, in: Proceedings of CPGPS 2010 Navigation and Location Services: Emerging Industry and International Exchanges, Scientific Research Publishing Inc., Shanghai, 2010.
[10] F.R. Helmert, Die Ausgleichungsrechnungnach der Methode der kleinsten Quadrate, Zweite Auflage, Teubner, Leipzig, 1907.
[11] Congwei Hu, Wu Chen, Yongqi Chen, Dajie Liu, Adaptive Kalman filtering for vehicle navigation, Journal of GPS 2 (1) (2003) 42–47.
[12] Karl-Rudolf Koch, Bayesian Inference With Geodetic Applications, Springer-Verlag, Berlin, 1990.
[13] Kun Qian, Jianguo Wang, Baoxin Hu, A posteriori estimation of stochastic models for multi-sensor integrated inertial kinematic positioning and navigation on basis of variance component estimation, Journal of GPS 2016 (14) (2016) 5.
[14] Kun Qian, Generic multisensor integration strategy and innovative error analysis for integrated navigation, York University, Canada, 2017.
[15] Kun Qian, Jianguo Wang, Baoxin Hu, Novel integration strategy for gnss-aided inertial integrated navigation, Geomatica 69 (2) (2015) 217–230.
[16] Kun Qian, Jianguo Wang, Nilesh Gopaul, Baoxin Hu, Low cost multisensor kinematic positioning and navigation system with Linux/RTAI, Journal of Sensors and Actuator Networks 1 (3) (2012) 166–182.
[17] Deren Li, Xiuxiao Yuan, Error Processing and Reliability Theory, Wuhan University Press, Wuhan, ISBN 7-307-03474-3, 2002.
[18] Deren Li, Ein Verfahren zur Aufdeckung grober Fehler mit Hilfe der a-posteriori-Varianz Schätzung, Bildmessung und Luftbildwesen 51 (5) (1993) 184–187.
[19] P. Matisko, V. Havlena, Noise covariance estimation for Kalman filter tuning using Bayesian approach and Monte Carlo, International Journal of Adaptive Control and Signal Processing 27 (11) (2013) 957–973.
[20] R.K. Mehra, On the identification of variances and adaptive Kalman filtering, IEEE Transactions on Automatic Control AC-15 (1970) 175–184.
[21] R. Mehra, Approaches to adaptive filtering, IEEE Transactions on Automatic Control 17 (5) (1972) 693–698.
[22] Ziqiang Ou, Estimation of variance and covariance components, Bulletin Géodésique 63 (1989) 139–148.
[23] Martin Salzmann, P.J.G. Teunissen, Quality Control in Kinematic Data Processing. Land Vehicle Navigation 1989, DGON Verlag TÜV Rheinland, Köln, 1989, pp. 355–366.
[24] H.W. Sorenson, Least-squares estimation: from Gauss to Kalman, IEEE Spectrum 7 (7) (1970) 63–68.
[25] Jianguo Wang, Filtermethoden zur fehler-toleranten kinematischen Positionsbestimmung, Schritenreihe Studiengang Vermessungswesen, Federal Arm-Forced University Munich, Germany, Neubiberg, 1997, No. 52.

[26] Jianguo Wang, Test statistics in Kalman filtering, Journal of GPS 7 (1) (2008) 81–90.
[27] Jianguo Wang, Reliability analysis in Kalman filtering, Journal of GPS 8 (1) (2009) 101–111.
[28] Jianguo Wang, Nilesh Gopaul, Bruno Scherzinger, Simplified algorithms of variance component estimation for static and kinematic GPS single point positioning, Journal of GPS 8 (1) (2009) 43–51.
[29] Jianguo Wang, Nilesh Gopaul, Jiming Guo, Adaptive Kalman filter based on posteriori variance–covariance component estimation, in: Proceedings of CPGPS 2010 Navigation and Location Services: Emerging Industry and International Exchanges, Scientific Research Publishing Inc., Shanghai, 2010.
[30] Jianguo Wang, Kun Qian, Baoxin Hu, An unconventional full tightly-coupled multi-sensor integration for kinematic positioning and navigation, in: CSNC 2015 Proceedings, Volume III, Vol. 7, No. 1, 2015, pp. 81–90.
[31] Yuanxi Yang, Weiguang Gao, An optimal adaptive Kalman filter, Journal of Geodesy 80 (4) (2006) 177–183.
[32] Zongchou Yu, ZhengLin Yu, Yanming Feng, The incorporation of variance component estimation in the filtering process in monitoring networks, Manuscripta Geodaetica 1988 (13) (1988) 290–295.
[33] Zongchou Yu, A universal formula of maximum likelihood estimation of variance-covariance components, Journal of Geodesy 70 (1996) 233–240.
[34] Ya Zhang, Jianguo Wang, Qian Sun, Wei Gao, Adaptive cubature Kalman filter based on the variance-covariance components estimation, Journal of Global Positioning Systems 2017 (15) (2017) 1–9.
[35] M. Stoehr, Der Kalman-Filter und seine Fehlerprozesse unter besonderer Berücksichtigung der Auswirkung von Modellfehlern, Forschung-Ausbildung-Weiterbildung Bericht Heft 19, Universität Kaiserlautern, Fachbereich Mathematik, August 1986.
[36] C.K. Chui, G. Chen, Kalman Filtering with Real-Time Applications, Fourth edition, Springer-Verlag, Berlin Heidelberg, ISBN 978-3-540-87848-3, 2009.
[37] Yongqi Chen, Jinling Wang, Reliability measures for correlated observations, Zeitschrift für Vermessungwesen 121 (5) (1996) 211–219.
[38] Hans Pelzer, Deformationsuntersuchungen auf der Basis kenematischer Bewegungsmodelle, Allgemeine Vermessungsnachrichten 94 (2) (1987) 49–62.
[39] Benzao Tao, Statistical Analysis of Surveying Data, Surveying and Mapping Press, Beijing, 1992.
[40] A.R. Amiri-Simlooei, Least-squares variance component estimation Theory and GPS Applications, PhD Dissertation, Delft University of Technology, 2007.
[41] W. Förstner, Ein Verfahren zur Schätzung von Varianz- und Kovarianzkomponenten, Allgemeine Vermessungsnachrichten 86 (11–12) (1979) 446–453.
[42] E. Grafarend, A. Kleusberg, B. Schaffrin, An introduction to the variance and covariance component estimation of Helmert type, ZfV 105 (4) (1980) 161–180.
[43] K.R. Koch, Maximum likelihood estimate of variance components, Bulletin Geodesique 60 (4) (1986) 329–338.
[44] P.J.G. Teunissen, A.R. Amiri-Simkooei, Least squares variance component estimation, Journal of Geodesy 82 (2) (2008) 65–82.
[45] Peiliang Xu, Yunzhong Shen, Yoichi Fukuda, Yumei Liu, Variance component estimation in linear inverse ill-posed models, Journal of Geodesy 80 (2) (2006) 69–81.
[46] Hermann Bähr, Zuheir Zuheir, Bernhard Heck, Variance Component Estimation for Combination of Terreatrial Reference Frames, Schriftenreihe des Studiengangs Geodäsie und Geoinformatik, Universität Karlsruhe, No. 6, 2007.
[47] Detlef Sieg, Milo Hirsch, Varianzkomponenten-schätzung in ingenieurgeodätischen Netzen, Allgemeine Vermessungsnachrichten 107 (3) (2000) 82–90.
[48] Volker Tesmer, Das stochastische Modell bei der VLBI-auswertung, PhD dissertation, No. 573, Reihe C, Deutsche Geodätische Kommission (DGK), Munich, 2004.

[49] Christian Tiberius, Frank Kenselaar, Variance component estimation and precise GPS positioning: case study, Journal of Surveying Engineering 129 (1) (2003).
[50] Andreas Rietdorf, Automatisierte Auswertung und Kalibrierung von scanneden Messsystemen mit tachy-metrischen Messprinzip, PhD dissertation, Civil Engineering and Applied Geosciences, Technical University Berlin, 2005.
[51] Jinling Wang, Chris Rizos, Stochastic assessment of GPS carrier phase measurements for precise static relative positioning, Journal of Geodesy 76 (2) (2002) 95–104.
[52] Alan S. Willsky, J.J. Deyst, B.S. Grawford, Adaptive filtering and self-test methods for failure detection and compensation, in: Proceedings of the 1974 JACC, Austin, Texas, June 19–21, 1974.
[53] Alan S. Willsky, J.J. Deyst, B.S. Grawford, Two self-test methods applied to an inertial system problem, Journal of Spacecraft 12 (7) (1975) 434–437.
[54] Alan S. Willsky, A survey of design methods for failure detection in dynamic systems, Automatica 12 (6) (1976) 601–611.
[55] R.K. Mehra, J. Peschon, An innovations approach to fault detection and diagnosis in dynamic systems, Automatica 16 (7) (1971) 637–640.
[56] Hermann Bähr, Zuheir Altamimi, Bernhard Heck, Variance Component Estimation for Combination of Terreatrial Reference Frames, Schriftenreihe des Studiengangs Geodäsie und Geoinformatik, Universität Karlsruhe, No. 6, 2007.

CHAPTER 4

Nonlinear control

Howard Li[a], Jin Wang[b], and Jun Meng[b]
[a]Department of Electrical and Computer Engineering, University of New Brunswick, Fredericton, NB, Canada
[b]Zhejiang University Robotics Institute, Hangzhou, China

Contents

Control theory deals with the behavior of dynamic systems with input, actuators, sensors, and output. A closed loop control system can modify the output by using feedback. The system to be controlled is called the process. The controller compares the output of the process with the desired output, and provides feedback to the process to modify the output in order to bring it closer to the desired output. Nonlinear control theory is concerned with the control theory that deals with systems that are nonlinear. In this chapter, linear system models and nonlinear system models are described. Then, some nonlinear control methods are introduced.

4.1 System modeling

A system can be modeled as linear systems, nonlinear systems, continuous systems, and discrete systems. Linear control systems are governed by linear differential equations while nonlinear control systems are governed by nonlinear differential equations.

Learning Control
https://doi.org/10.1016/B978-0-12-822314-7.00009-2

4.1.1 Linear systems

A linear system is described as

$$\dot{\mathbf{x}}(t) = A\mathbf{x}(t) + B\mathbf{u}(t),$$
$$\mathbf{y}(t) = C\mathbf{x}(t) + D\mathbf{u}(t), \tag{4.1}$$

where $t \in R$ denotes time, $\mathbf{x} \in R^{n\times 1}$ is the state, $\mathbf{u} \in R^{m\times 1}$ is the input or control, $\mathbf{y} \in R^{p\times 1}$ is the output, $A \in R^{n\times n}$ is the dynamics matrix, $B \in R^{n\times m}$ is the input matrix, $C \in R^{p\times n}$ is the output or sensor matrix, $D \in R^{p\times m}$ is the feedthrough matrix.

A discrete linear system is described as

$$\mathbf{x}_t = A\mathbf{x}_{t-1} + B\mathbf{u}_{t-1},$$
$$\mathbf{y}_t = C\mathbf{x}_t, \tag{4.2}$$

where the subscript t denotes time.

4.1.2 Nonlinear systems

A nonlinear system is described as

$$\dot{\mathbf{x}}(t) = f(\mathbf{x}, \mathbf{u}, t),$$
$$\mathbf{y}(t) = g(\mathbf{x}, \mathbf{u}, t), \tag{4.3}$$

where $f(\cdot)$ and $g(\cdot)$ represent the nonlinear system dynamics and nonlinear sensor measurement, respectively.

A discrete nonlinear system is described as

$$\mathbf{x}_t = f(\mathbf{x}_{t-1}, \mathbf{u}_{t-1}),$$
$$\mathbf{y}_t = g(\mathbf{x}_t). \tag{4.4}$$

4.2 Nonlinear control

Nonlinear control theory deals with real-world control systems that are nonlinear (governed by nonlinear differential equations). Some classical nonlinear control methods are described.

4.2.1 Feedback linearization

We should not expect to be able to cancel nonlinearities in every nonlinear system. There must be a certain structural property of the system that allows us to perform such cancellation.

The ability to use feedback to convert a nonlinear state equation into a controllable linear state equation by [1] canceling nonlinearities requires the nonlinear state equation to have the structure

$$\dot{\mathbf{x}} = A\mathbf{x} + B\gamma(\mathbf{x})[\mathbf{u} - \alpha(\mathbf{x})] \tag{4.5}$$

where $\alpha(\mathbf{x})$ can be canceled by subtraction and $\gamma(\mathbf{x})$ can be canceled by division, respectively.

We can linearize the state equation via the state feedback control law

$$\mathbf{u} = \alpha(\mathbf{x}) + \beta(\mathbf{x})\mathbf{v} \tag{4.6}$$

to obtain the linear state equation

$$\dot{\mathbf{x}} = A\mathbf{x} + B\mathbf{v} \tag{4.7}$$

where $\beta(\mathbf{x}) = \frac{1}{\gamma(\mathbf{x})}$.

For stabilization, we design

$$\mathbf{v} = -K\mathbf{x} \tag{4.8}$$

such that $A - BK$ is Hurwitz, i.e. there are poles on the left hand side to make the system stable. The overall nonlinear stabilizing state feedback control is

$$\mathbf{u} = \alpha(\mathbf{x}) - \beta(\mathbf{x})K\mathbf{x} \tag{4.9}$$

where $\beta(\mathbf{x}) = \frac{1}{\gamma(\mathbf{x})}$.

Example. Consider the system

$$\dot{x} = ax^2 - x^3 + u, \tag{4.10}$$

feedback linearization can be used to stabilize x at $x = 0$.

The overall nonlinear stabilizing state feedback control is

$$u = -(ax^2 - x^3) - kx. \tag{4.11}$$

Then,

$$\dot{x} = ax^2 - x^3 + u = -kx, \tag{4.12}$$

$$\dot{x} + kx = 0. \tag{4.13}$$

Therefore, the system is stable.

4.2.2 Stability and Lyapunov stability

Before nonlinear controllers are introduced, the stability of a system has to be defined.

Let $\mathbf{x} = 0$ be an equilibrium point for

$$\dot{\mathbf{x}} = f(\mathbf{x}) \tag{4.14}$$

and $D \subset R^n$ be a domain containing $\mathbf{x} = 0$.

Lyapunov Stability Theorem. *Let $V : D \to R$ be a continuously differentiable function such that $V(0) = 0$ and $V(\mathbf{x}) > 0$ in $D - \{0\}$, $\dot{V}(\mathbf{x}) \leq 0$ in D. The derivative of $V(\mathbf{x})$ along the trajectories of $\dot{\mathbf{x}} = f(\mathbf{x})$, denoted by $\dot{V}(\mathbf{x})$, is given by*

$$\begin{aligned} \dot{V}(\mathbf{x}) &= \Sigma_{i=1}^{n} \frac{\partial V}{\partial x_i} \dot{x}_i \\ &= \Sigma_{i=1}^{n} \frac{\partial V}{\partial x_i} f_i(\mathbf{x}) \\ &= \frac{\partial V}{\partial x} f(\mathbf{x}). \end{aligned} \tag{4.15}$$

Then, $\mathbf{x} = 0$ is stable. *Moreover, if $\dot{V}(\mathbf{x}) < 0$ in $D - \{0\}$, then $\mathbf{x} = 0$ is* asymptotically stable.

Example. Consider the system

$$\dot{x} = ax^2 - x^3 + u. \tag{4.16}$$

Define the Lyapunov function

$$V = \frac{1}{2}x^2. \tag{4.17}$$

The control law is

$$u = -ax^2 - kx. \tag{4.18}$$

Then

$$\dot{V} = x\dot{x} = x(ax^2 - x^3 + u) = x(-x^3 - kx) = -(x^4 + kx^2) < 0. \tag{4.19}$$

Therefore, the system is asymptotically stable.

4.2.3 Sliding mode control

Consider the second-order system

$$\dot{x}_1 = x_2, \tag{4.20}$$

$$\dot{x}_2 = h(\mathbf{x}) + g(\mathbf{x})u, \tag{4.21}$$

where h and g are unknown nonlinear functions and $g(\mathbf{x}) \geq g_0 > 0$ for all $\mathbf{x}$. We want to design a state feedback control law u to stabilize the origin $\mathbf{x} = \mathbf{0}$.

The sliding mode control [2] consists of a reaching phase during which trajectories starting off the manifold $s = 0$ move toward it and reach it in finite time, followed by a sliding phase during which the motion is confined to the manifold $s = 0$ and the dynamics of the system are represented by the reduced-order model

$$\dot{x}_1 = -a_1 x_1. \tag{4.22}$$

The manifold $s = 0$ is called the sliding manifold and the control law u is called sliding mode control (Fig. 4.1).

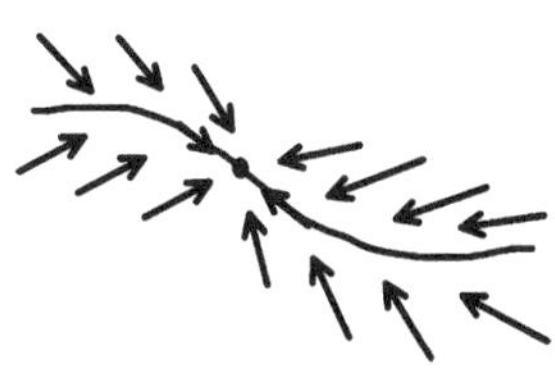

Figure 4.1 Sliding mode control.

Example.

Step 1: Design a sliding surface

$$s(x_1, x_2) = a_1 x_1 + x_2 = 0. \tag{4.23}$$

Step 2: Design a controller to bring the system to the sliding surface

$$s(x_1, x_2) = 0. \tag{4.24}$$

In order to bring the variable $s = a_1 x_1 + x_2$ to 0, we need to find a Lyapunov candidate function $V(s)$ that satisfies the Lyapunov criterion. Once $s = 0$ (system on the sliding surface), the system will become stable because $\dot{x}_1 = -a_1 x_1$.

Choose

$$V(s) = \frac{1}{2}s^2 \geq 0 \tag{4.25}$$

as a Lyapunov function candidate where

$$\dot{s} = a_1\dot{x}_1 + \dot{x}_2 = a_1x_2 + h(\mathbf{x}) + g(\mathbf{x})u \tag{4.26}$$

To be able to find the control law u, it is necessary that the nonlinearity is bounded. Recall $g(\mathbf{x}) > 0$. Suppose $\forall \mathbf{x} \in R^2$ for some known function $\rho(\mathbf{x})$, h and g satisfy the inequality

$$|\frac{a_1x_2 + h(\mathbf{x})}{g(\mathbf{x})}| \leq \rho(\mathbf{x}), \tag{4.27}$$

$$\begin{aligned}
\dot{V}(s) &= s\dot{s} \\
&= s[a_1x_2 + h(\mathbf{x}) + g(\mathbf{x})u] \\
&= s[a_1x_2 + h(\mathbf{x})] + g(\mathbf{x})su \\
&\leq g(\mathbf{x})|s|\rho(\mathbf{x}) + g(\mathbf{x})su \\
&= g(\mathbf{x})(|s|\rho(\mathbf{x}) + su).
\end{aligned} \tag{4.28}$$

Choose

$$u = -\beta(\mathbf{x})\text{sign}(s) \tag{4.29}$$

where $u = -\beta(\mathbf{x})\text{sign}(s)$ is used only for $s \neq 0$ since in ideal sliding mode control u is not required on the sliding surface $s = 0$; $\beta(\mathbf{x}) > \rho(\mathbf{x})$ (the control needs to be more negative than the upper bound of the nonlinearity $\rho(\mathbf{x})$); and $\text{sign}(x) = 1(s > 0), 0(s = 0), -1(s < 0)$; and this yields

$$\begin{aligned}
\dot{V} &\leq g(\mathbf{x})(|s|\rho(\mathbf{x}) - s\beta(\mathbf{x})\text{sign}(s)) \\
&< g(\mathbf{x})|s| \times 0.
\end{aligned} \tag{4.30}$$

Therefore, the trajectory reaches the manifold $s = 0$ in finite time.

4.2.4 Backstepping control

Consider the system

$$\begin{aligned}
\dot{\mathbf{z}} &= f(\mathbf{z}) + g(\mathbf{z})\xi, \\
\dot{\xi} &= \mathbf{u},
\end{aligned} \tag{4.31}$$

where $\mathbf{z}$ cannot be controlled by $\mathbf{u}$ directly.

A controller can be developed to control one derivative at a time [3].

Step 1: Find the virtual control law $\xi = \phi(\mathbf{z})$ that can stabilize $\mathbf{z}$ under the ideal condition

$$\dot{\mathbf{z}} = f(\mathbf{z}) + g(\mathbf{z})\xi = f(\mathbf{z}) + g(\mathbf{z})\phi(\mathbf{z}). \tag{4.32}$$

Step 2: Minimize the difference between ξ and the ideal virtual control law $\phi(\mathbf{z})$ to stabilize the error $\epsilon = \xi - \phi(\mathbf{z})$.

Step 3: Combine the control for both $\mathbf{z}$ and ϵ for the overall system. Find the control $\mathbf{u}$ to stabilize $\mathbf{z}$ and ξ.

Example.

Step 1: Suppose we can use ξ as the control input to asymptotically stabilize the first half of the system $\dot{\mathbf{z}} = f(\mathbf{z}) + g(\mathbf{z})\xi$ with a control law

$$\xi = \phi(\mathbf{z}) \tag{4.33}$$

where $\phi(\mathbf{z})$ is the virtual control, and $\phi(\mathbf{z} = 0) = 0$. This implies that the origin $\mathbf{z} = 0$ of

$$\dot{\mathbf{z}} = f(\mathbf{z}) + g(\mathbf{z})\phi(\mathbf{z}) \tag{4.34}$$

is asymptotically stable.

Suppose also we know a Lyapunov function $V(\mathbf{z})$ for $\mathbf{z}$ which satisfies the inequality

$$\frac{\partial V(\mathbf{z})}{\partial \mathbf{z}}\dot{\mathbf{z}} = \frac{\partial V}{\partial \mathbf{z}}[f(\mathbf{z}) + g(\mathbf{z})\phi(\mathbf{z})] \leq -W(\mathbf{z}) \tag{4.35}$$

where $W(\mathbf{z})$ is positive definite.

Step 2: In reality $\xi \neq \phi(\mathbf{z})$, the difference between the state variable ξ and the virtual control $\phi(\mathbf{z})$ is

$$\epsilon = \xi - \phi(\mathbf{z}) \tag{4.36}$$

Therefore

$$\xi = \phi(\mathbf{z}) + \epsilon. \tag{4.37}$$

With the tracking error ϵ, in reality the system (4.31) is

$$\begin{aligned}\dot{\mathbf{z}} &= f(\mathbf{z}) + g(\mathbf{z})\xi \\ &= f(\mathbf{z}) + g(\mathbf{z})[\phi(\mathbf{z}) + \epsilon]\end{aligned}$$

$$
\begin{aligned}
&= [f(\mathbf{z}) + g(\mathbf{z})\phi(\mathbf{z})] + g(\mathbf{z})\epsilon, \\
\dot{\xi} &= \mathbf{u}.
\end{aligned} \tag{4.38}
$$

In order to stabilize $\mathbf{z}$, we need to make the state variable ξ track the virtual control $\phi(\mathbf{z})$ by minimizing the error $\epsilon = 0$.

From (4.38), we have

$$
\begin{aligned}
\dot{\mathbf{z}} &= [f(\mathbf{z}) + g(\mathbf{z})\phi(\mathbf{z})] + g(\mathbf{z})\epsilon, \\
\dot{\epsilon} &= \dot{\xi} - \dot{\phi}(\mathbf{z}) = \mathbf{u} - \dot{\phi}(\mathbf{z}).
\end{aligned} \tag{4.39}
$$

Since $\mathbf{z}$ is controlled by the virtual control $\phi(\mathbf{z})$, the control problem becomes to find $\mathbf{u}$ to stabilize ϵ (minimize $\epsilon = 0$) so that the second half of the system is also stable.

A Lyapunov function is defined to stabilize ϵ,

$$
V(\epsilon) = \frac{1}{2}\epsilon^2. \tag{4.40}
$$

Step 3: Combine the Lyapunov functions

$$
V_c(\mathbf{z}, \epsilon) = V(\mathbf{z}) + V(\epsilon). \tag{4.41}
$$

From (4.39),

$$
\begin{aligned}
\dot{V}_c(\mathbf{z}, \epsilon) &= \frac{\partial V}{\partial \mathbf{z}}\dot{\mathbf{z}} + \frac{d}{dt}(\frac{1}{2}\epsilon^2) \\
&= \frac{\partial V}{\partial \mathbf{z}}[f(\mathbf{z}) + g(\mathbf{z})\phi(\mathbf{z}) + g(\mathbf{z})\epsilon] + \epsilon\dot{\epsilon} \\
&= \frac{\partial V}{\partial \mathbf{z}}[f(\mathbf{z}) + g(\mathbf{z})\phi(\mathbf{z})] + \frac{\partial V}{\partial \mathbf{z}}g(\mathbf{z})\epsilon + \epsilon(\mathbf{u} - \dot{\phi}(\mathbf{z})) \\
&\leq -W(\mathbf{z}) + \frac{\partial V}{\partial \mathbf{z}}g(\mathbf{z})\epsilon + \epsilon(\mathbf{u} - \dot{\phi}(\mathbf{z})).
\end{aligned} \tag{4.42}
$$

Choose

$$
\mathbf{u} = \dot{\phi}(\mathbf{z}) - \frac{\partial V}{\partial \mathbf{z}}g(\mathbf{z}) - k\epsilon \tag{4.43}
$$

where $k > 0$. Therefore

$$
\dot{V}_c(\mathbf{z}, \epsilon) \leq -W(\mathbf{z}) - k\epsilon^2, \tag{4.44}
$$

which implies the origin of (4.39) is asymptotically stable ($\mathbf{z} = 0$, $\epsilon = 0$). Therefore $\xi = \phi(\mathbf{z}) + \epsilon = 0$ because $\phi(\mathbf{z} = 0) = 0$. This implies the origin of the original system (4.31) is stabilized by $\mathbf{u}$.

Note: $\dot{\mathbf{z}}$ in (4.39) is different from the ideal case in Step 1 because $\xi \neq \phi(\mathbf{z})$. $\dot{\mathbf{z}}$ in (4.39) includes the virtual control error ϵ. The derivative of $V_c(\mathbf{z}, \epsilon)$ is along the trajectory defined by (4.39).

4.2.5 Adaptive control

According to the Webster's dictionary, to adapt means: "to adjust oneself to particular conditions; to bring oneself in harmony with a particular environment; to bring one's acts, behavior in harmony with a particular environment".

If the system model can be identified continuously, then we can compensate variations in the model of the system simply by varying adjustable parameters of the controller and thereby obtain a satisfactory system performance continuously under various conditions [4].

Example. For system

$$\dot{x} = ax + bu, \tag{4.45}$$

a and b can be known or unknown.

1) a and b known, linear feedback control:
The control law is

$$u = -kx \tag{4.46}$$

where $|k| > |a/b|$ to have a control that is large enough.

2) a and b unknown, adaptive control:
If a and b are unknown and $b > 0$, consider the feedback control law

$$u = -kx \tag{4.47}$$

with adaptive k in which

$$\dot{k} = x^2 \tag{4.48}$$

so that k is adaptive in order to increase the feedback control. Therefore,

$$\begin{aligned} \dot{x} &= ax - bkx, \\ \dot{k} &= x^2. \end{aligned} \tag{4.49}$$

Consider the positive Lyapunov function

$$V(x, k) = \frac{1}{2}(x^2 + b(k - \hat{k})^2) \tag{4.50}$$

for $\hat{k} > 0$. Then

$$\begin{aligned}\frac{dV}{dt} &= x\dot{x} + b(k - \hat{k})\dot{k} \\ &= x(ax - bkx) + b(k - \hat{k})x^2 \\ &= -x^2(\hat{k}b - a) \\ &\leq 0 \end{aligned} \tag{4.51}$$

provided we pick $\hat{k}$ such that $\hat{k}b > a$.

Furthermore,

$$\frac{dV(x, k)}{dt} = 0 \tag{4.52}$$

when $x = 0$ for arbitrary k.

Therefore, the system is stable.

4.3 Summary

In this chapter, system models are introduced. Nonlinear control techniques are described, including feedback linearization, Lyapunov stability, sliding mode control, backstepping, and adaptive control. Examples are presented to illustrate nonlinear control techniques.

References

[1] R.W. Brockett, Feedback invariants for nonlinear systems, IFAC Proceedings Volumes 11 (1) (1978) 1115–1120.
[2] V.I. Utkin, Variable structure systems with sliding modes, IEEE Transactions on Automatic Control AC22 (1977) 212–222.
[3] P.V. Kokotovic, The joy of feedback: nonlinear and adaptive, IEEE Control Systems Magazine 12 (3) (1992) 7–17.
[4] I.D. Landau, Adaptive Control: The Model Reference Approach, Marcel Dekker, 1979.

CHAPTER 5

Deep learning approaches in face analysis

Duygu Cakir, Simge Akay, and Nafiz Arica
Computer Engineering Department, Bahcesehir University, Istanbul, Turkey

Contents

5.1 Introduction

A scene can tell a lot about a face such as where it is located, where it is looking at, what it has, how it is looking, and who it belongs to. One of the most interesting yet complex and unresolved problems in computer vision is the automatic face analysis which can answer these questions. The aim is to extract as much information as possible to use in many fields one can think of. Naming a few would be to extract the information of location, pose, gender, ID, age, expression, and to use in systems such as automation [228],

Learning Control
https://doi.org/10.1016/B978-0-12-822314-7.00010-9

medicine [113,169,73], distance education [70,8], biometrics [125,3,16], security [173,90,109,75] and forensics [31,63], robotics [103,106], authentication [58,16,40], driver monitoring [170,50,60], surveillance [122,107], and even mobile phone apps [97,162] such as aging, putting on make-up, changing the style, measuring the attractiveness, etc. Although the human eye and brain can catch major or minor differences like occlusion, pose, illumination, expression, aging, beard, change in hair style, and makeup, the computer vision area is not robust to all these changes and it is still lacking the ability to interpret them all.

The automatic recognition of faces, attributes and facial expressions requires robust face detection and tracking systems. By the late 1980s and early 1990s, cheap computing power started becoming available, hence led to the development of face detection and tracking algorithms [12]. Data labeling, analysis and processing problems have been solved, and regular / irregular relations between data are uncovered faster and more accurately through time with the help of the improvement of powerful processing units, storage space, and data collection. In addition to those improvements, the developments of new algorithms in neural network training result with a deep learning approach, a family of machine learning methods. Deep learning, inspired by the human brain, has improved the accuracy of many artificial intelligence tasks including face analysis. The most common deep networks are Convolutional Neural Networks, Recurrent Neural Networks, Reinforcement Learning and recently Generative Adversarial Networks.

Convolutional Neural Networks (CNNs) were first introduced by Fukushima [62] which were then improved by LeCun [101] who used CNNs for digit recognition. A more simplified and generalized version was introduced by Behnke [10] and Simard et al. [180] in 2003. Although it was developed for document recognition, since they reduce the complexity, CNNs are used in many facial expression recognition studies such as [99]—being the first study on facial expression recognition using CNNs, [137], [64]. In 2006, with the increased efficiency on GPU computing, since it became more possible to train large networks, CNNs were introduced with multi-layer studies such as [82,134,35,146].

Recurrent Neural Networks (RNNs) were first proposed by Rumelhart et al. [176] to analyze temporal sequences of inputs such as sentences, videos, speech, etc. They can store representations of recent input events in the form of activations. Long short-term memory (LSTM) [83] is a special type of RNN which is designed to learn to store information over extended term intervals. The first application to images was in 2009 [72] for handwriting recognition. One of the most important applications of an LSTM is an automatic image caption generator [198], where natural sentences are generated to describe the image / scene. LSTMs are also used for many face related tasks such as face recognition and anti-spoofing [104,214,171], face reconstruction [236,49], facial attribute detection [80], expression recognition [209,183,223].

A generative model to represent and implement game theory in a whole new form, called Generative Adversarial Networks (GANs) has been proposed as a pioneer study by Goodfellow et al. [69]. GANs are a way to generate new data and they are proven to work in the form of unsupervised learning, as well as semi-supervised and supervised learning. GANs are usually used for face recognition [192,41] and face synthesis [84,6] such that they are authentic to human observers. They are also generally used in face aging methods [206,4,124], face generation from textual description [228,26] or from voice [207] for popular applications. Another disputable application area is face applications in which a face is mapped to another person's head [145]. There are many ethical debates and legal restrictions going on in the field of face manipulation.

One last method worth mentioning would be Reinforcement Learning (RL). Unlike supervised or unsupervised learning methods, RL does not need marked input training / testing set labels, on the contrary, it tries to find a balance between exploration and exploitation by trial and error in a dynamic environment to be able to control data, usually using a method called Markov Decision. It is generally used in implementing artificial game players and one of the earliest and successful examples would be TD-Gammon [190]. Although RL was not originally generated as a deep learning method, in 1997, DeepMind converted the algorithm to a deep perspective and implemented the first commercial application: Deep Blue, and afterwards came AlphaGo. Besides game development, RL is usually used in face localization [112], face recognition [168,205], and preprocessing [24,151].

A face analysis system starts with the detection of face regions in the scene. Given an image or image sequence, the detection algorithm tries to find and localize the face or faces. This problem can be considered as a different version of object detection which only aims for the face and nothing else, not its identity, not its pose, nor its attributes. After face detection, which produces bounding boxes each containing a single face image, pre-processing algorithms are applied in order to prepare and im-

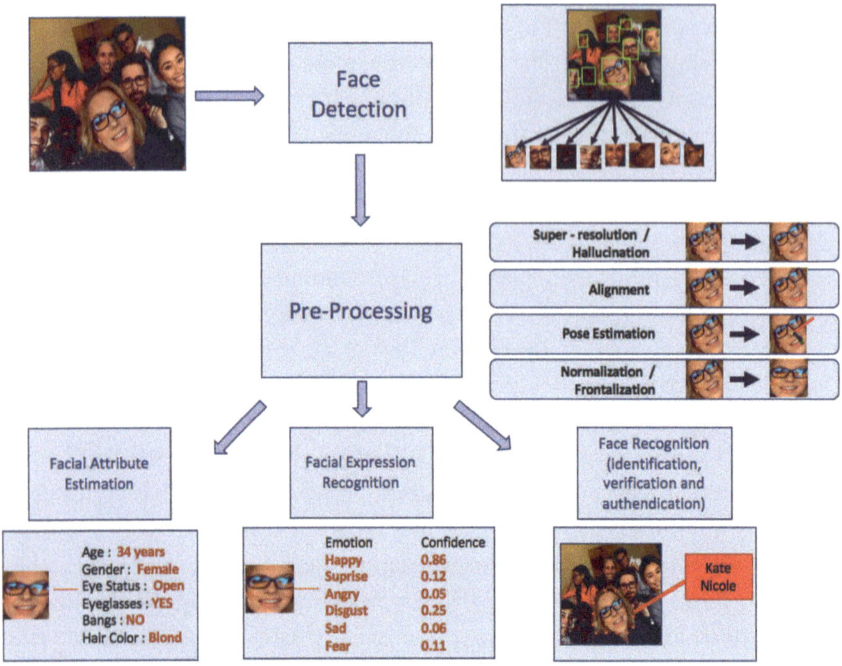

Figure 5.1 The conventional pipeline in face analysis.

prove the representative attributes of face image for the subsequent stages. Which pre-processing algorithms are to be employed differs based on the goal of the analysis. Although they are related and overlapped with each other, they will be discussed in four groups, namely frontalization (normalization), super resolution (hallucination), alignment and pose estimation. While face alignment identifies the geometric structure of human faces by facial landmarks, pose estimation determines the head pose by estimating roll, pitch and yaw angles. Face frontalization aims to synthesize a frontal view of input face image.

After the pre-processing stage, face analysis proceeds with various algorithms including facial attribute estimation, facial expression recognition, face identification and verification. Facial attributes can be considered as the mid-level semantic features of the face such as age, gender, having glasses, and blonde hair to describe an input face. Facial expression recognition tries to label a face with one of the emotional states, including happiness, sadness, surprise, fear, anger, disgust and neutrality. Face recognition contains three subtasks: identification, verification, and authentication.

This chapter covers all the stages of the face analysis pipeline including face detection, pre-processing, facial attribute estimation, expression classification, and face recognition. The outline is visually summarized in Fig. 5.1. It should be noted that the pipeline does not necessarily have to follow these steps in the exact same order. Some algorithms or estimation steps can be merged to increase accuracy or to serve the required problem, nevertheless, the overall order stays the same. The remainder of this chapter will provide a structured presentation of deep state-of-the-art techniques applied to face analysis in computer vision, mainly over the last decade, followed by the related issues and main challenges in the field. Lastly, a conclusion will be provided to sum up the overall study and a guide will be suggested for the future works.

5.2 Face detection

Face detection can be regarded as a particular case of object detection where it ignores anything else except facial features by determining the location of human faces. There are a lot of challenges in face detection, reducing detection accuracy. These challenges include complex background with a lot of objects in the image, multiple faces, illumination, low-resolution, skin color, head pose, facial occlusions such as glasses, hands, etc. Various studies have been performed to provide improvements in the literature. Recently, a number of deep learning methods have been developed and demonstrated for face detection [166]. In general, face detection methods can be grouped into three categories, namely sliding window, region proposal, and single-shot approaches.

5.2.1 Sliding window

Sliding window approaches compute a detection score and bounding box coordinates at every location in a feature map. Although it does not employ a deep neural network it is worth mentioning the Viola-Jones face detection algorithm [199] because it is still the most popular method which achieves high detection rates in an effective way. The algorithm has four stages; rectangular feature selection, creating an integral image, training, and cascade of classifiers. In Fig. 5.2.a, an example rectangular features displayed relative to the enclosing detection window is given. For the fast computation, an integral image is created by computing values at each pixel (x, y) that is the sum of the pixel values above and to the left of (x, y) (see

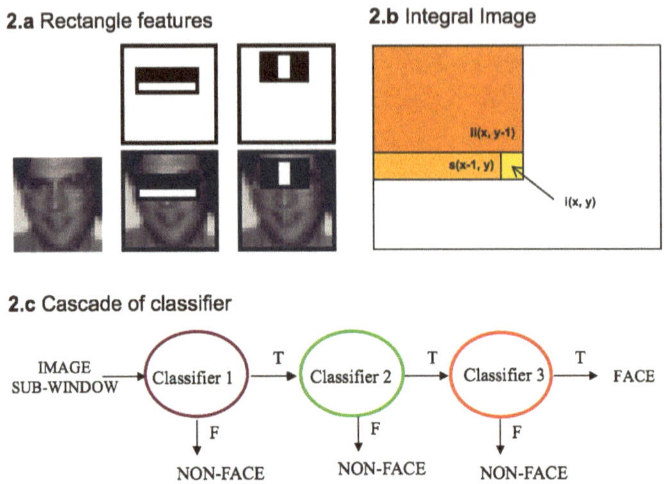

Figure 5.2 The Viola-Jones face detection algorithm [199].

Fig. 5.2.b). The algorithm employs the AdaBoosting classifier [59] by combining a set of weak classifiers in each iteration within the framework of an attentional cascade. As shown in Fig. 5.2.c, a cascade of classifiers is applied to each sub-window. Overall, the Viola-Jones is the most widely used and efficient algorithm for detecting faces, but its accuracy decreases in detecting faces in-the-wild. To solve this problem, many studies have applied features like HOG [245], SIFT [128], SURF [9], ACF [215], and some heuristic approaches [155,19]. More recently, deep learning methods have been deployed with noteworthy results [108,229].

Sliding window-based deep learning approaches include Deep Dense Face Detector (DDFD) [55], Deep Pyramid Deformable Part Model (DPDPM) [164], and Faceness-Net [217]. In all these methods, the networks produce face-part responses, along with full face responses, and combine them based on their spatial configuration to determine the final face score. DDFD uses a single model based on deep CNNs without using pose or facial landmark annotation. It uses the advantage of the capacity of deep CNNs to detect faces from multiple views and minimize the computational complexity by simplifying the architecture.

In another study, HOG features are replaced with features learned by a fully convolutional network [57]. This novelty provides to generate a pyramid of deep features, analogous to a HOG feature pyramid that is called the full model Deep Pyramid DPM [68]. The normalization layer is added to

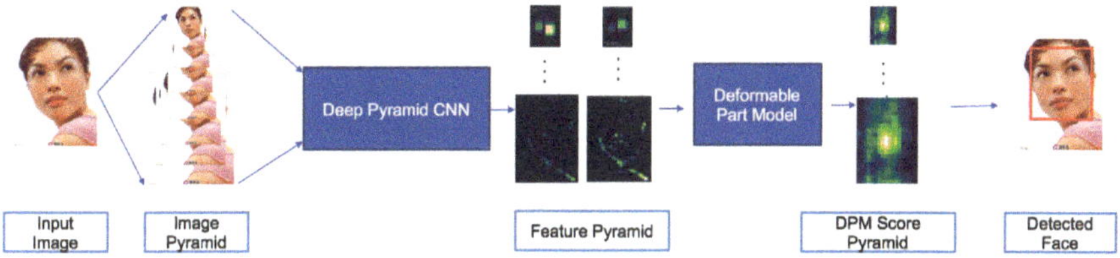

Figure 5.3 Overview of Deep Pyramid Deformable Part Model (DP2MFD) [164].

this architecture by Ranjan et al. [164] and named DP2MFD. Fundamentally, DP2MFD consists of two models. The first one uses a normalized deep feature pyramid, which extracts the fixed-length features from each location using sliding windows. The second one uses the SVM classifier to determine each location as a face or not (Fig. 5.3). The feature activations in CNN diversify extensively in each layer of the feature pyramid. To reduce the bias to face size that occurred from this diversity, a normalization layer is added to the CNN, and root filter DMP is obtained using a linear SVM.

A recent study Faceness-Net [217], uses the facial attribute-based supervision for learning a face detector. Faceness-Net gives face-likeness scores based on deep network responses on local facial parts. In the first stage, it applies multiple attribute aware networks to generate response maps of different facial parts such as eyes, hair, nose, mouth, and beard. In the second stage, it fixes candidate windows that are generated from the first stage using CNN, where face classification and regression are jointly optimized. It computes a faceness score on these generated windows [217]. If the score is higher, they are chosen as the top candidate windows.

5.2.2 Region proposal

Region proposal approaches basically select face regions independently from the input image and compute a hyper-dimensional feature vector to classify whether or not a given proposal contains a face. Examples of successful region-based DNN approaches are Faster R-CNN [172], Selective Search [196], Hyperface [165], and All-in-one Face [167].

Girshick et al. [67] proposed R-CNN (Region with CNN features), where the selective search is applied to extract fewer regions from the image instead of trying to classify a huge number of regions. It gives better results, but it still takes a tremendous amount of time and region proposal to train. To solve these problems, the same author of the paper (R-CNN) proposed a

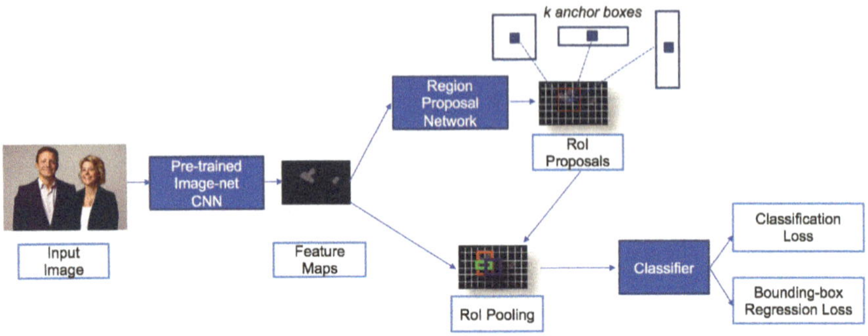

Figure 5.4 The architecture of Faster R-CNN.

new and updated approach called Fast R-CNN [66] to build a faster detection. From the convolutional feature map, a region of proposals is warped by using a Region Of Interest (ROI) pooling layer; then they are reshaped into a fixed size so that it can be fed into a fully connected layer. Both algorithms (R-CNN and Fast R-CNN) utilize a selective search to determine the region proposals. Selective search is a slow and time-consuming algorithm that adversely affects performance. Therefore, a Faster R-CNN has been proposed [172], which eliminates the selective search algorithm. Instead of a selective search, a separate network is used to predict the region proposals (Fig. 5.4). It consists of a regional proposal network (RPN) and a fast R-CNN detector network. RPN gives a set of rectangles based on a deep convolutional layer. Mostly, RPN slides a little window (3×3) on the feature map to label the window as a face or not, and furthermore gives a bounding box area.

HyperFace [165] is another approach using the selective search algorithm in R-CNN to create region suggestions for faces in the image. It is a multi-task deep convolutional method which consists of three modules. The first module produces a region proposal class, independent from a given image, and scales it. In the second module, CNN takes the resized candidate regions and classifies them as face or non-face. The last module is a post-processing step, which involves iterative region proposals to increase the face detection score and improve the performance. A new network called All-in-one Face has been developed by Ranjan et al. [167], that gives significantly better results than HyperFace. Like Hyper-Face, it employs a multi-task learning framework, but it benefits from regularization and network initialization.

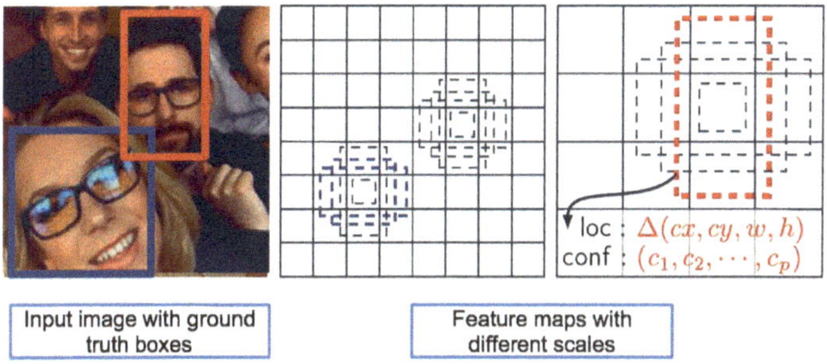

Figure 5.5 Single Shot Detector (SSD) Framework.

5.2.3 Single shot

Single Shot approaches apply a single forward algorithm. These approaches take one single shot to detect multiple faces within the image and evaluate a set of default boxes at each location in several feature maps with different scales. According to some recent surveys [166,136], few single shot approaches have been proposed recently. Liu et al. [119] proposed a method named SSD (Single Shot Detector) for detecting objects in images using a single deep neural network. It evaluates a set of default boxes at each location in several feature maps with different scales. As shown in Fig. 5.5, SSD only needs an input image and ground truth boxes for each object during training. A small set of default boxes of different aspect ratios with different scales are evaluated in a convolutional network. Similar to Faster R-CNN, it regresses to offsets for the center (c_x; c_y) of the default bounding box and for its width (w) and height (h). The shape offsets and the confidences for all object categories are predicted for each default box ($c_1, c_2, \ldots, c_p$).

Another approach, ScaleFace [218], detects faces of different scales from multiple layers of the network, and these networks are finally incorporated. It consists of a region proposal network and a Fast R-CNN classifier. Fast R-CNN has got two convolutional layers and two fully connected layers, besides, the region proposal network has got one convolutional layer. Furthermore, Zhang et al. [233] proposed S3FD, which uses a scale equitable framework and scale compensation anchor matching strategy for improved detection of small faces.

It is clear that face detection is the most critical part of the face analysis system. Face detection algorithms return the coordination of the bounding boxes containing the faces in the input image. In the next step, the face images are prepared for the subsequent high level analysis.

5.3 Pre-processing steps

Pre-processing is one of the most crucial steps in deep learning-based face analysis problems. All input images should be normalized into a standard form to make the deep learning model robust in terms of visual deformations. At an end-to-end face analysis system, pre-processing begins with face alignment, which is generally achieved by using the facial landmarks on the detected face. One or more processing steps are then applied based on the requirements of the subsequent analysis problem. This section summarizes face pre-processing steps under four topics: face alignment, head pose estimation, face frontalization, face super resolution.

5.3.1 Face alignment

Face alignment identifies the geometric structure of human faces. It attempts to obtain a canonical alignment of the face based on translation, scale, and rotation using facial landmarks such as eye centers, mouth corners, and the tip of the nose [92,18,150,5]. Most of the face alignment algorithms use these landmarks to provide either by discriminative fitting or cascaded regression. There are some clear benefits to align the face, such as establishing correspondence among different faces so that the subsequent image analysis tasks can be performed on a common basis. Face alignment, which scales and rotates through the correct facial points, needs the power of deep learning due to some challenges in low-level features.

5.3.1.1 Discriminative model fitting

Discriminative fitting-based approaches generally use a holistic template to fit the facial shape for a given input facial image. Examples of discriminative fitting-based approaches are Active Appearance Model [36], Active Shape Model [197], Constrained Local Model [224]. These approaches learn the features to handle two-dimensional shapes.

Besides 2D shape-based approaches, 3D model-based face alignment methods have been recently developed. Jourabloo et al. (2016) proposed a 3D approach which employs a cascaded CNN-based regression for estimating the coefficients of a 3D to 2D matrix. The authors then improve their work by formulating the face alignment problem as a dense 3D model-fitting, where the camera and 3D shape parameters are predicted (Fig. 5.6) [87]. Given a set of training faces and their augmented representations, the cascaded CNN coupled-regressor estimates the desired update of parameter in every stage of CNN. 3D Dense Face Alignment (3DDFA) method by

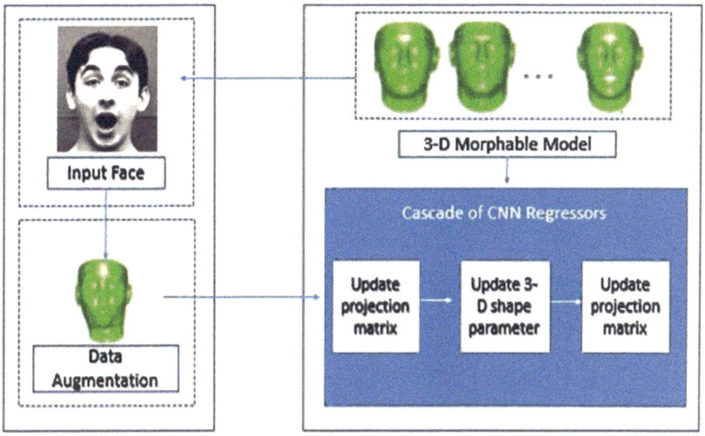

Figure 5.6 The overall process of CNN-based 3D model fitting [87].

Zhu et al. [246] fits a dense 3D face model to the image via CNN, where the depth data is modeled. Different from traditional methods, 3DDFA leaves out the 2D landmarks and starts from 3D fitting with cascaded CNN.

5.3.1.2 Cascaded regression

Cascaded regression-based approaches learn the features by refining the predicted shape progressively. Cao et al. [25] designed an explicit shape regression method that uses feature selection. Xiong and De la Torre [213] proposed a supervised descent method which refines the facial shapes learning feature mapping. Nevertheless, the features obtained in these methods are hand-crafted, that are ineffectual to represent local patches. Therefore, deep learning approaches have been utilized to extract more robust features for attaining high accuracy.

Sun et al. [186] propose a deep cascaded CNN method which consists of an initialization and a refinement stage. Outputs of multiple networks are fused for landmark estimation at each level to achieve excellent performance. DCNN has individual progress for each landmark but it is delicate for former prediction. To address this, several approaches [231,96] have been proposed to fuse the feature for predicting facial landmarks. Zhang et al. [231] proposed a coarse to fine autoencoder networks (CFAN), which cascades several successive stacked autoencoder networks. Lai et al. [96] presented a deep cascaded regression method (DCRFA) under the deconvolutional neural networks. Trigeorgis et al. [193] employed a combined and jointly trained convolutional recurrent neural network architecture that

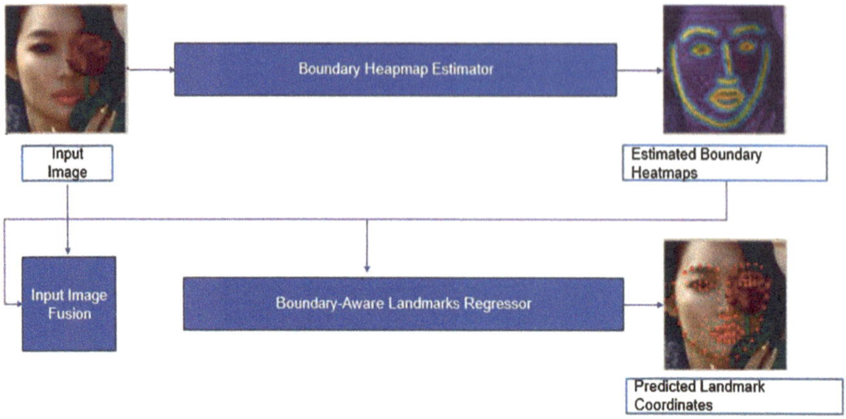

Figure 5.7 Boundary-Aware Face Alignment framework [211].

enables training of the regressors jointly used in the cascaded regression framework called mnemonic descent method (MDM). More recently, Wu et al. [210] proposed a boundary aware landmark alignment algorithm that increases the localization accuracy by facial boundaries. Their algorithm firstly estimates the boundary heatmap, then it incorporates boundary information into feature learning. After the boundary heatmap fusion, it generates the final prediction of landmarks (Fig. 5.7).

5.3.2 Pose estimation

Head pose is a 3D vector containing the yaw, pitch and roll angles of a given face image. Estimating the head pose essentially requires it to learn a mapping between 2D and 3D spaces. It can be used to improve expression recognition [226], identity recognition [192], attention detection [33], etc.

While some approaches use facial landmarks to estimate head pose [91,20], some others utilize more modalities such as 3D information or temporal information [141,143,135,54]. Facial landmarks-based methods mainly use deep neural networks for pose estimation [246,220,154]. Zhu et al. [246] present a 3D Morphable Model (3DMM) [14], which achieves face alignment across large poses by minimizing the difference between image and model appearance. 3DMM, containing yaw, pitch and roll parameters is fitted into the input image by a cascaded CNN. The method solves the self-occlusion in modeling and the high nonlinearity in fitting when large poses exist.

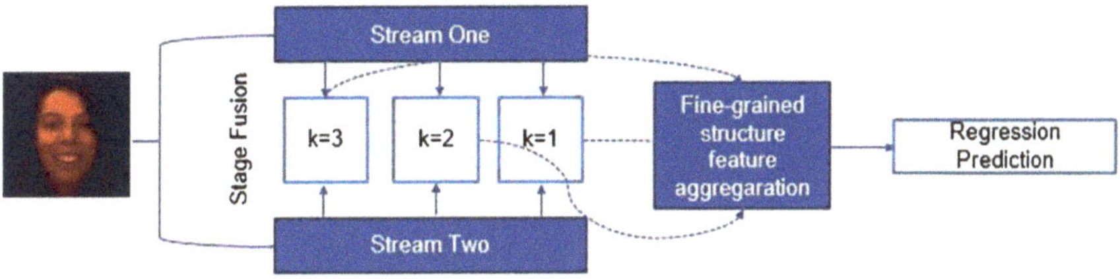

Figure 5.8 Fine-Grained Aggregation Network (FSA-Net) framework [219].

A more recent study, Fine-Grained Aggregation Network (FSA-Net) [219] uses direct regression without facial landmarks. As shown in Fig. 5.8, it consists of two stages. One is spatially group pixel-level features of the feature map together and uses them for fusion. The second one is learning to find the fine-grained structure mapping for spatially grouping pixel-level. For each feature map, the network computes its attention map, then fine-grained structure mapping learns to extract pixel-level features in the feature maps.

Despite the considerable attention on face alignment, pose estimation researches are rather limited in the literature. Several approaches try to perform different related facial analysis tasks together. Hyperface [165] can perform face detection, facial landmark localization and head pose estimation using CNN features. KEPLER [94] learns local features by a heatmap CNN.

5.3.3 Face frontalization

Face frontalization is the pre-process of synthesizing a frontal view of a face, given its non-frontal view. Face frontalization methods can be grouped into two categories, namely 2D/3D local texture warping and Generative Adversarial Networks-based methods.

5.3.3.1 2D/3D local texture warping

A good representative study of frontalization process based on 2D/3D local texture warping is proposed in [79]. It defines pose normalization as the process of back-projecting the appearance of an input face image to the reference coordinate system using the 3D surface as a proxy. The approach is developed using a non-personalized, pre-established 3D model to approximate the camera matrix. Based on the camera matrix, the query face is re-shaped into a frontal one, using bilinear interpolation. The method uses five main steps; namely landmark detection, pose estimation, frontal pose

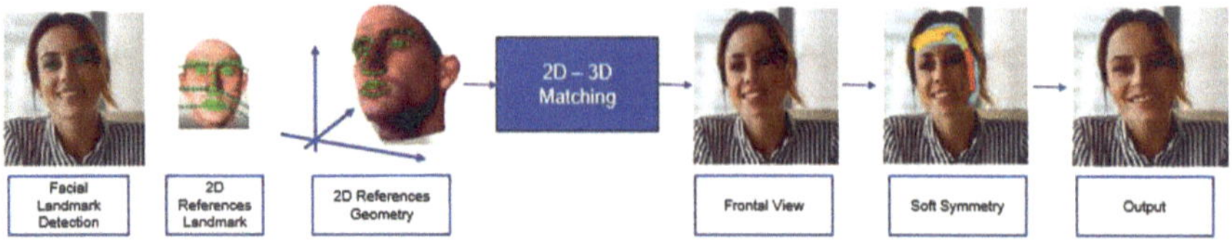

Figure 5.9 Architectures for 2D/3D local texture [79].

generation, visibility estimation, and mirroring. After landmark detection, the camera matrix is estimated to capture the input image by exploiting the correlation between landmarks in the input image and 3D landmarks in the reference model. Pixels of the input image are then back projected, creating the frontal posed face image. It constructs a visibility map on the semi-normalized face by counting overlapping projected pixels. In the next step the invisible parts are filled by the mirror of visible parts on the other side to generate a more visually satisfying output. Fig. 5.9 shows the architecture in [79]. This study is improved in a recent study by a Stacked Denoising Auto-Encoder (SDAE) in [27]. Instead of applying a mirroring operation for the invisible face parts of the posed image, SDAE learns how to fill in those regions by a large set of training samples.

5.3.3.2 Generative adversarial networks (GAN) based

In recent years, GAN, has been successfully introduced into the field of face frontalization. It was firstly used by DR-GAN [192] which takes one image as the input, termed single-image, and the extended model that leverages multiple images per subject, termed multi-image. Later, TP-GAN [84] was proposed to model the synthesis function with one single network. It has two pathways; one global network and four landmarks-located patch network. It mainly aggregates local textures in addition to the commonly used global encoder-decoder network. CR-GAN [191] is proposed using two-pathway; leveraging unlabeled data and generating high-quality multi view images.

All those methods treat face frontalization as a 2D image translation problem. Recently, in-the-wild databases have increased to consider 3D properties of the human face. FF-Gan [221] combines knowledge of 3D face using generative adversarial networks, which consists of a generator and a discriminator network. The generator takes a non-frontal face as input to generate a frontal output, while discriminator attempts to classify it as a real frontal image or a generated one. In an FF-Gan network, first, the reconstruction module predicts 3D Model coefficients. 3D Model coefficients support pose and frequency information. Then the generator tries to synthesize realistic-looking images that can fool distinguish a real image from a synthetic one. In the final step, the discriminator separates generated faces against real ones.

5.3.4 Face super resolution

Super-resolution imaging (SR) is a technique to increase the resolution of an input image. In Face SR, global model-based approaches perform low-resolution (LR) image on the whole face, often by a learned mapping between LR and high-resolution (HR) face images. Liu and Sun [114] proposed a Bayesian inference, which is a two-step statistical inference model that integrates both a global parametric linear model and a local nonparametric one. It exploits evidence to update the state of the uncertainty over competing probability models. Part-based approaches handle the facial parts separately and design a patch-based reconstruction model in the high-dimensional kernel space. Nhan Duong et al. [147] proposed a model that divides the face image into four parts, which are eyes, mouth, nose, and cheek; reconstructing each area separately.

State-of-the-art face resolution methods try to explore and utilize image domain priors based on deep learning. Cootes et al. [37] propose a deep CNN model that predicts a residual image. It uses a cascade of layers to transform into HR. Cootes et al. present an attention aware face resolution approach that specifies an attended region and enhance it by considering the global perspective of the whole face at each step [36]. In contrast to other approaches that often learn a single patch mapping from LR to HR images, the attention aware approach takes to deep reinforcement learning sequentially, then performs the facial part recovery by fully exploiting the global mutuality of the given image. At each step, the recurrent policy network takes face resolution results and encodes them as an action history vector by LSTM [65] as the input, and then action probabilities outputs are kept into a probability map. In the final stage, the algorithm extracts the local patch given the probability map and generates the final enhanced patch. A more recent study face super resolution network (FSR-Net) [32] utilizes the facial points to resolve LR face images without any alignment requirement.

5.4 Facial attribute estimation

Attributes are labels that can be assigned to an image to describe its appearance and its mid-level semantic features. Rather than focusing on the identity, the attributes have been made the core of image recognition problem [56] to describe an image. An image can contain one or more attributes depending on the scene and attribute type. The concept of attributes for face verification has been introduced by Kumar et al. [95] to shift the task of recognition from *naming* to *describing*. As being the first group bringing up the concept of *attributes*, [95] attempts to verify faces under controlled settings using attributes. After their ground-breaking method, many researchers continued to use attribute recognition for tasks such as facial expression analysis, and gender recognition. Some of these studies use region-based algorithms [11], and some use multi-task representation learning to examine the latent relationships [52] or hypothetical correlations [21].

As in many computer vision problems, deep learning has also been applied in attribute detection. Studies show that facial attribute detection techniques have been basically divided into two sub-categories; 1: locating the region-of-interest of the desired attribute and 2: modeling the relationships between attributes. From this point on, this section will investigate the attribute detection techniques under these two topics which mainly focus on localizing the Region of Interest (ROI) in which the attribute takes place and modeling the relationships to detect attributes.

5.4.1 Localizing the ROI

Locating the positions of facial attributes helps the feature extractors and attribute classifiers to focus more on attribute-relevant regions / patches. If it is to find if "eyeglasses" exist or not, or to find an Action Unit (AU) leading to an expression classification, the main goal in these types of studies is to locate the areas where the attribute may take place. Then a classifier is applied to the ROI to detect the corresponding attribute.

One of the first studies using deep learning was the "Poselet" [17], in which given a whole person image, the detector finds the appearance and the configuration of the body parts. However, the detector only examines the body image and pose of the desired body part rather than the details of the face. A poselet-based algorithm on facial attributes has been proposed in [89] by taking the whole image as an input and augments a CNN to have input layers based on semantically aligned part patches. The model learns features that are specific to a certain part under a certain pose, which forms the poselets. These poselets help to factor out the pose and viewpoint variation which allows the convolutional network to focus on the pose-normalized appearance differences.

Another study [115] proposes a deep learning framework which uses two CNNs for attribute prediction in-the-wild. While LNet localizes the face regions by averaging response maps of attribute filters, ANet extracts the high-level face representation to predict the attributes without alignment. The two CNNs are fine-tuned jointly with attribute tags but pre-trained differently. LNet uses the entire image for training to localize the face without the bounding boxes or landmarks and without normalization. The weakly supervised manner of LNet makes data preparation easier since only image level attribute tags of training images are provided with a massive amount of object categories to have a good generalization capability. Then ANet extracts discriminative face representation, making attribute prediction from the entire face region possible, again pre-trained with massive face identities and is fine-tuned by attributes. While LNet uses the entire face region rather than focusing on the relevant part, [46] proposes a model in which the model learns the relevant regions simultaneously in a cascade network to make the classification.

Amongst the approaches that have been mentioned above and many others [131,133,81], despite the fact that they argue that it is the most proper way to focus on the region which triggers the attribute or in which the attribute takes place, the region-based models' accuracies rely too much on correct face detection or localization. So far, it has been important to localize the attributes, but rather than relying on independent classifiers, it is also crucial, and intuitive, to describe the shared information and latent correlation among each other as well as attribute-specific information for more discriminative features.

5.4.2 Modeling the relationships

Looking at the popular datasets like LFW or CelebA, one can easily spot that attributes, such as bald and child or baby and round jaw, may be strongly related and not independent at all. For an improved classification, a relationship model has to be built between these attributes. The current methods model the attribute relationships with the help of prior, human-labeled knowledge; hence it is also a challenge to build the latent relationships automatically and adaptively, and optimize them jointly. Through this relationship modeling, the model can eavesdrop, i.e. learn task

A through task B [175], using the *hints* [2] it gains by predicting the most important features of the tasks.

One of the first methods that use multi-task learning to model the relationship between attributes is [1] in which they use multiple CNNs, each for predicting one binary attribute and generate attribute-specific feature representations. Then they apply a multi-task learning to show that attributes in the same group share more knowledge and attributes in other groups generally compete each other, and consequently share less knowledge. Another study [174] addresses the multi-label imbalance problem by introducing a novel mixed objective optimization network (MOON). MOON takes the classification as a regression problem and predicts the scores of multiple attributes to reduce the regression error, and show that a multi-objective approach is more effective than an independent classifier.

In a related study, a multi-task deep convolutional neural network (MCNN) with an auxiliary network at the top (AUX) named MCNN-AUX is proposed. Their model takes advantage of high-level attribute relationships (categories) for improved classification in three ways: 1. by sharing the lowest layer of for all attributes, 2. by sharing the higher layers for spatially-related attributes, and 3. by feeding the attribute scores from MCNN into the AUX network to find score-level relationships. Moreover, their work does not use pre-training, which may have improved their results to a far higher level. They suggest that by taking advantage of implicit and explicit relationships among attributes will lead to improved facial recognition. Based on the model MCNN, [23] designed a partially shared structure called PS-MCNN. Unlike MCNN, which groups the attributes into categories by their task-specific (or semantic) similarities, PS-MCNN groups them into four categories according to their regions. Similar to this model, [76] proposes a categorization system based on type/scale (ordinal vs. nominal) and semantic meaning of the attributes, using which they learn the shared features of all the attributes and category-specific features of heterogeneous attributes.

A recent study, [241], a bi-directional network, which they named BLAN, models to learn hierarchical representations, covering the correlations between feature hierarchies and attribute characteristics. In their study, they derive a loss to further incorporate the locality of facial attributes to high-level representations at each hierarchy. The proposed BLAN architecture can be found in Fig. 5.10.

To name a few more, [244,129,235,148,241] are works that build up CNNs to demonstrate the relationships between attributes to fortify at-

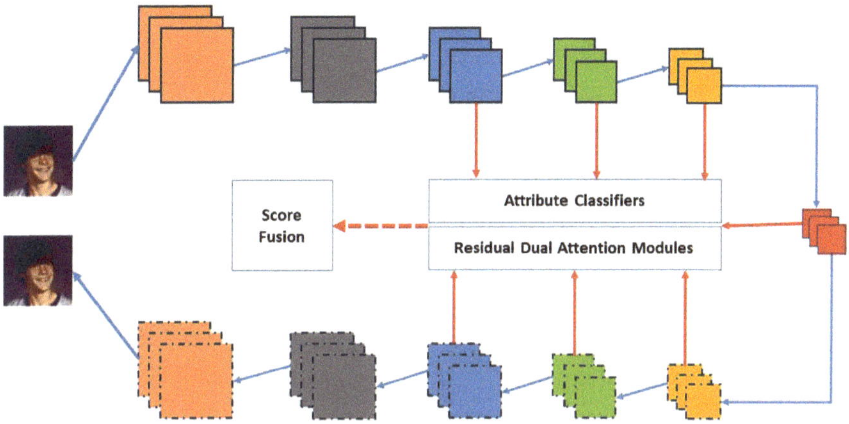

Figure 5.10 The overall architecture of the Bi-directional Network (BLAN) [241].

tribute classification among many other facial estimation tasks. Examining the recent studies, although it is a topic belonging to another research, it can easily be said that the attribute classification trend has been moving towards Graph Neural Networks [71], in which they learn a target node's representation by exploring neighbor information in an iterative manner until a stable node is reached. The main issue on these relationship modeling models is that the discovery of the relationships still relies too much on human-touch, the categorization of the attributes is manually done by the researchers themselves. Discovering the relationships adaptively, without prior information, should be the focus of future studies [240].

5.5 Facial expression recognition

It is stated that during a typical face-to-face communication, verbal components convey only one-third of the social meaning whereas two-thirds of the social meaning comes from non-verbal components such as body gestures and facial expressions [13]. Moreover, Mehrabian [139] claims that facial expressions constitute up to 55% of a conversation. Facial expressions have been studied by clinical and social psychologists, medical practitioners, actors and artists. However, in the last quarter of the 20th century, with the advances in the fields of robotics, computer graphics, and computer vision, computer scientists started showing interest in the study of facial expressions [12]. Even though the first study goes back to 1978 [248] in which the system uses 20 tracking points, there has been almost no study until the beginning of the 1990s.

After the study of Charles Darwin [39], which lists the general principles of expressions in humans and animals, the second milestone is the study on facial muscles and their movements for each expression in the work of Ekman [61]. Their work has a significant influence on today's modern facial expression analysis studies. According to the study of Ortony and Turner [149], regardless of culture, there are 6 universal emotions: Happiness, Sadness, Surprise, Fear, Anger and Disgust; 7 if the neutral face is included. Since these are claimed to be universal, they are usually correctly-classified by many studies. Ekman and Friesen [53] describe each basic emotion in terms of a facial expression called Action Units (AU) that uniquely characterizes that emotion. Recently, [86] states that the 6 basic emotions are culture-specific and not universal. [194] suggests a new approach to the emotional states and calls it the 2D valence-arousal emotion space in which they investigate moods rather than emotions. There are certainly many other emotions than the six above. The six emotions are known as the basic or prototypic emotions, whereas in 2001 Parrott identified 136 emotions which can be categorized in separate classes and subclasses [153]. Another noteworthy research is [7] where 400 facial expressions are studied.

The aim in Facial Expression Recognition (FER) is to obtain discriminative characteristics of the face through the geometric, appearance, and hybrid features using this extreme power. Facial expressions provide information about the affective state as well as the cognitive activity, temperament and personality, truthfulness and psychopathology of a human being [48]. Having a wide set of application areas, FER has become popular as a commercial detection and a marketing agent. Besides the studies on single static images, there are many more that have been studying the temporal analysis of facial expressions in image sequences, i.e. video input [239,216,44,243,34,181], with the availability of the increased capacities of computing power. Although the human eye can catch minor expressions, the computer vision area is still lacking the ability to recognize and interpret those expressions. The higher gain in the accuracy of these systems will lead to offline analyzer tools as well as real-time analyzers in which the system will be able to socially interact with the people and analyze their expressions.

As the other face related tasks, most FER approaches generally follow the same set of stages; after the face is detected, we have a feature extraction algorithm or a few algorithms fused together, followed by a suitable classification method. Some studies use low level feature extraction methods and their combinations for FER [238,120,21,226], whereas some use high level ones [102,182,159].

Although the techniques developed for FER are enhanced in many areas, it is still quite a challenge to interpret the emotions in real time since the in-the-wild (ITW) settings have illumination, pose and noise differences, and most important of all, occlusion. Having eyeglasses or a beard is quite a common occlusion, so it is investigated enough, and the success rate is satisfactory, but having a huge scar, bangs, or even an eye-patch are uncommon attributes which may distort the expression and affect the FER task. This is mainly the result of having insufficiently labeled data under many occluded settings. Considering these issues, deep neural networks are proven to be successful and robust towards these types of face analysis tasks [64], moreover, they do not need low-level feature sets, they can learn features automatically by the nature of the network. In short, deep networks can execute FER in an end-to-end manner.

FER studies are problem-specific, they highly depend on the input, and vary in many ways. Some (actually most) work with static images in which the feature representation is encoded with only spatial information from the current frame [179,118,142], some with image sequences (videos) considering the temporal relation among frames [88,237], some only use simple networks [142], and some implement a multi-task network to jointly train multiple networks to learn the interaction of FER and another task [140,157,247] or a cascaded network to sequentially train multiple networks in a hierarchical approach to strengthening the learned features and eliminate the others [186,132,117]. Some even study multi-modal relationships such as audio and psychological channels [38]. Since there are many works focusing on gaining the same output using different methods, this research focuses on the most important works implemented in the field rather than categorizing them into sub-sections.

In their work, Mavani et al. [138] present a novel technique of visual saliency using intensity maps to show the field of attention using AlexNet [93]. After fine-tuning the network, a softmax layer of seven outputs is applied. Four out of the six emotions are successfully classified across two datasets. Another successful study using a cross-database method is [142], in which they use two convolution layers, both followed by max pooling and then four inception layers, which is known to have used the inception layer for the first time across datasets. Other CNN-based studies are [222] which uses multiple CNNs; [116] which works across datasets; [85] which collects depth data and feeds it to the CNN.

The temporal relations extracted from consecutive frames in a video sequence using 3D convolutional networks and Long Short-Term Memory (LSTM) have been studied by Hasani and Mahoor [77] to improve the recognition of subtle changes in the facial expressions in a sequence by extracting the facial landmarks. Another study [234] on temporal sequences uses a Recurrent Neural Network (RNN) by modeling the facial morphological variations and the evolutionary properties of expression, which is said to be effective to capture the dynamic variation of the facial physical structure. Ebrahimi et al. [51] also mentions a hybrid of CNN and RNN systems using temporal averaging for aggregation.

Though the researchers mainly focus on landmarks, [28] presents a method for modeling 3D face shape, viewpoint, and expression from a single, unconstrained photo using image intensities without the use of facial landmarks. Their method uses three deep CNNs to estimate each of these components separately and work as a stand-alone landmark detector. Unlike the methods that use the information of facial landmarks, their algorithm estimates these properties directly from image intensities. They test their method in two datasets: lab-controlled CK+ [130] and in-the-wild EmotiW [43]. They take the peak frames of CK+ and all the frames of EmotiW with multiple versions of the datasets with different scales. Surprisingly, they state that older landmark detectors work better than the newer ones and furthermore say that better landmark detection accuracy does not necessarily translate to better face pre-processing. Other approaches that use expression intensity rather than detecting landmarks for expression recognition are [105] and [187].

Research to model the high-level neurons of an expression network called FaceNet2ExpNet is proposed by Ding et al. [47] in a two stage algorithm to be able to work on small datasets. In the pre-training stage, they train the convolutional layers of the expression net, to regularize a FER network by using a probabilistic distribution function for the modeling of high level neuron responses. In their refining stage, they append the randomly initialized fully-connected (FC) layers to the pre-trained convolutional layers and train the whole network jointly with cross-entropy loss. In their first stage, the regression loss is back propagated to the ExpNet to learn the rich face information from FaceNet. In the second stage, they attach the FC layer and train the model only with the expression labels to increase the discriminative capability (Fig. 5.11). Their FER accuracies cannot compete

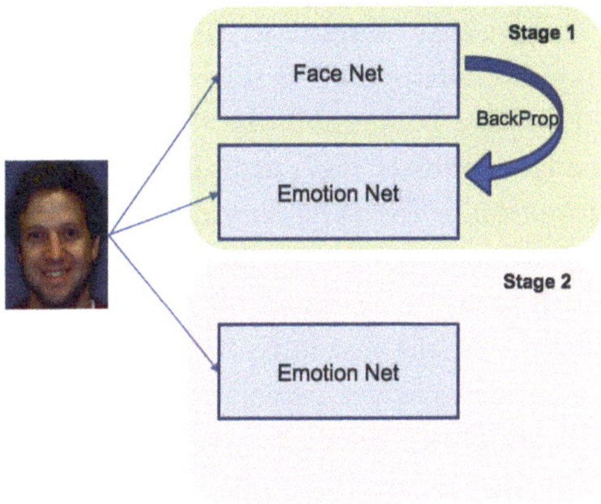

Figure 5.11 Two stage training algorithm of FaceNet2ExpNet [47].

with models that use multiple CNN layers but they are successful amongst their opponents.

It is stated that using a neural network comes with its trade-offs [200]. They suggest that, if it is a frontal (or some-what clear) face, the CNN needs fewer layers and hyper-parameters; however, if it has recorded ITW, more layers and parameters are needed. One must find a balance between the number of layers and parameters for occluded yet clear faces.

5.6 Face recognition

Recognizing a person is one of the most important issues in the field of computer vision. Whether it is a commercial business or a single researcher, finding the identity of a subject is under the radar of many fields, mainly including surveillance, security, and forensics as well as entertainment. The face is the physical part that holds the most important discriminative features for person recognition, especially when the world is moving towards a digital age where we shop online using digital money instead of paper.

Face recognition (FR) contains three sub-tasks: identification, verification, and authentication. Face identification computes 1:N similarity between the dataset and the image searching for a face in the dataset of many faces to find a match (Who are you?). Face verification computes 1:1 similarity, trying to compare the image one by one to find a match (Is that you?). Face authentication tries to find the match and find the access status (Are you allowed to...?). This part of the survey will not focus on these subtasks one by one since they all relate to each other, instead, it will focus on the deep techniques implemented to solve the issues in FR.

As Masi et al. [136] are concerned, the first study that points out the importance of Face Recognition (FR) dates back in 1966 [15], where the researcher draws attention to the variability in head rotation, tilt, lighting, facial expression, aging, etc., and furthermore says that the experiments fail to find the correlation between 2 pictures of the same person with different head rotations. Which is still the case today... 25 years after the first research, the experiments continued using a newer technique which was going to be still used today, called the eigenface, where they apply linear algebra for face identification [195]. After the significant jump in neural networks using backpropagation to recognize handwritten digits [100] (which was enhanced in [101] and used CNN for the first time in images), researchers have moved the shift of data analysis from nominal data to images with the implementation of AlexNet, which uses deep learning [93] and made great breakthrough experiments, and after that, starting with DeepFace with the data collected from Facebook [189] and FaceNet with the data collected from Google [177], FR made a big leap using deep neural networks. The timeline of the deep FR model milestones [204] can be found in Table 5.1.

In 2014, Facebook developed DeepFace [189] which uses CNN to extract features from 4 million images out of more than 4000 subjects. Using the implementation power combined with the data they hold in hand, they implemented the network with nine layers with two convolutions. DeepFace aligns the face images to a 3D model and obtains a face representation. This work demonstrates that employing an analytical 3D model alignment based on fiducial points improves the accuracy of state-of-the-art models significantly.

Sun et al. [184] show that FR can be well solved by using both face identification and verification signals as supervision. The Deep IDentification-verification features (DeepID2) are learned with CNNs using four layers, where the first three use max pooling. Since ReLU has better fitting abilities than the sigmoid units for large training datasets [93], they use ReLU for neurons in the convolutional layers and the DeepID2 layer. In their verification task, they detect 21 facial landmarks, align the images by similarity transformation according to the extracted landmarks, crop 400 patches, and create 400 DeepID2 vectors extracted by a total of 200 CNNs, each of which is trained to extract two 160-dimensional DeepID2 vectors on one particular face patch and its horizontally flipped counterpart, respectively, of each face. Then they learn the Joint Bayesian

Table 5.1 Different verification methods with their architecture and loss functions.

Method	Loss	Architecture
DeepFace [189]	Softmax	AlexNet
DeepID2 [184]	Contrastive loss	AlexNet
DeepID3 [185]	Contrastive loss	VGGNet-10
FaceNet [177]	Triplet loss	GoogleNet-24
Baidu [126]	Triplet loss	CNN-9
VGGface [152]	Triplet loss	VGGNet-16
light-CNN [211]	Softmax	Light CNN
Center Loss [208]	Center loss	Lenet +-7
Range Loss [230]	Range loss	VGGNet-16
L2-softmax [163]	L2-softmax	ResNet-101
Normface [202]	Contrastive loss	ResNet-28
CoCo loss [123]	CoCo loss	–
vMF loss [78]	vMF loss	ResNet-27
Marginal Loss [42]	Marginal loss	ResNet-27
SphereFace [121]	Softmax	ResNet-64
CCL [160]	Center invariant loss	ResNet-27
AMS loss [201]	AMS loss	ResNet-20
Cosface [203]	Cosface	ResNet-64
Arcface [41]	Arcface	ResNet-100
Ring loss [242]	Ring loss	ResNet-64

model [30] for face verification based on the extracted DeepID2. They state that FR is to develop effective feature representations for reducing intra-personal variations while enlarging inter-personal differences and the key for a successful FR is to increase inter-personal variations and reduce intra-personal variations. The face identification task increases the inter-personal variations by drawing DeepID2 extracted from different identities apart, while the face verification task reduces the intra-personal variations by pulling DeepID2 extracted from the same identity together.

The following year, another group from Google implemented FaceNet [177] with the data collected from Google. For pre-processing, their model directly learns a mapping from face images to a compact Euclidean space where the squared L2 distances correspond to face similarity, which indicates lower distances for the same person. Afterwards a deep CNN is trained using triplets of roughly aligned matching / non-matching face patches. Their approach learns its representation from a large dataset of labeled faces and directly from their pixels to achieve robustness for illumination, pose, and other variational conditions. While doing this, they employ two architectures [225,188] to design an end-to-end system; where they use a batch input layer and a deep CNN, followed by L2 normalization, and finalized with a triplet loss. Their model strength lies in not using a CNN bottleneck layer by directly optimizing the embedding itself, and having almost no pre-processing other than cropping the face.

Meanwhile, on the east coast of the US, [123] focus on the features and propose an angular softmax loss that enables CNN to learn angularly discriminative features, SphereFace, by defining a large angular margin (A-Softmax loss), which leads to the constrained region to become smaller, and the learning task to be more difficult. While the most popular loss functions (as the one mentioned above) employ the Euclidean margin to the learned features, their method directly considers angular margin since it links directly to discriminativeness on a manifold.

There are many more studies worth mentioning, but since there is a limited space, the above-mentioned methods will be wrapped up here. For more detailed deep face recognition analysis, one can check the following survey papers [158,136,29].

5.7 Discussion and conclusion

Deep neural networks have shown to perform very well on large scale object recognition problems and simple ones. Even so, since face analysis became a study of focus, other issues have emerged. This section will cover the main issues mentioned even in the earliest researches as well as recent issues that came out with the advancement in computing.

Pose

Having a non-frontal head pose is assumed as a self-occlusion [161]. Most of the systems are modeled to recognize the frontal image. Although deep systems can overcome this issue, it is still an issue that creates loss of information on the face. [110] suggests that 3D face models can cope with this problem. Pose invariant deep models are also suggested by many researchers [45].

Illumination

Illumination is one of the most common problems in all areas of computer vision. Although there are some low-level pre-processing techniques to overcome the issue (logarithmic transforms, histogram equalization, gamma intensity correction) or high-level techniques [212,227], a robust solution has yet to be found. Since the pre-processing techniques will only be able to improve the problem, but not heal it totally, the collected data type is starting to change. The fusion of other modalities, such as infrared images, depth information from 3D face models and physiological data, is becoming a promising research direction [110].

Occlusion

Partial occlusions on the face can occur due to facial hair, hands covering the face, accessories such as glasses, scarves, veil, external objects that occlude the camera (a mobile phone held in hand in front of the face), etc. Occlusion leads to the loss of informative facial features and it is one of the most challenging problems in face related tasks. Kotsia et al. [156] claims that occlusion of the mouth leads to inaccuracies in recognition of anger, fear, happiness and sadness, whereas the occlusion of the eyes decreases the accuracy in recognition of disgust and surprise. There is no specific solution for occlusion, but one way seems to be the reconstruction of the occluded parts for static images. One other solution would be to follow the prior knowledge of the image sequence in the temporal data to predict the occluded part. Reconstruction will not work if the occluded part is a whole and non-symmetrical body part (like the lower face being totally occluded) and prior knowledge will not function since expressions are instantaneous.

Lack of data

In FER research, there are lots of datasets but most of them are acted (frontal), and most of them are static images. Very few of them offer in-the-wild (ITW) data, in which the subjects are usually movie characters meaning that it is still an acted data. If the subject becomes aware that (s)he is being recorded, the expression of the face loses its authenticity [12]. Even though movie scenes are considered to be ITW, the actors highly exaggerate the emotion, in some cases, making the level of expression artificially exaggerated. Even though enough ITW data is found / collected, it is hard and time consuming to label the data. In some cases (like Action Unit labeling), it also requires some expertise of the attribute coder. Some systems try to automatically label the dataset (like a fully automatic crowd-sourcing algorithm), yet it still needs human assistance with an expert eye. Some suggest that considering the environment may help to automatize the labeling. The researchers gave a splendid example for this: If there is a birthday party, the happy context has the highest weight and the outliers can be ignored [22].

Another issue in data is that in some datasets, some labeled classes have many more samples than some others [74]. Even if it is the most used dataset, CK+ has this issue by having 18 contempt and 83 surprise labels, also having 2 AU13 and 287 AU25 labels! Although it can be negligible if data augmentation is applied to the minor labels for FER studies, when it comes to analyzing the attributes, the face cannot be augmented without another attribute being triggered. Also, besides the six so-called-universal emotions, very few data can be provided and very little work seems to have been done. There is also the problem of not being able to combine datasets in the same study. Since almost all datasets are structured differently, very few cross-dataset studies are available.

Overfitting

Training the data too well may happen in a perfect environment, but real data is not perfect, not even close. Everything seems to be working excellent, the training accuracy may become closer and closer to 100%, but the minute you try to test your network, the system crashes since the model memorizes everything in the training data. The deep network has to have some noise in order to not memorize the training data.

ITW images require more layers to learn complex shapes. The more layers, the more hyper-parameters. This can easily be discarded by randomly dropping the neurons, which also helps the overfitting problem. Another solution to get rid of the overfitting problem is to add more layers to the model, i.e. increase the complexity, or changing the fine-tuning, or increasing the training data. Some other solutions would be early stopping the network, pruning, cross-validation, regularization, and data augmentation [111,127,217].

Expressions

The presence of expressions and their level (strength, intensity) is a huge struggle for FR. Researchers have come up with approaches to dealing with this issue like model, muscle, and motion-based approaches [144]; however, none of them seems to be working perfectly. There is also the problem of blended expressions.

Subjectivity

Especially FER is a person-dependent task, and the subject may have blended expressions. Subjects may combine the typical expression with another AU or may be unable to express feelings due to many reasons such as diseases or psychological disorders.

Examining the results in FER studies, there are some expressions usually confused with others, which seems natural. For instance, anger is usually confused with disgust and fear is confused with surprise [178]. However, although there are studies supporting the similarity between anger and disgust, the confusion between fear and surprise was not very evident. It has been shown that out of the seven expressions, surprise and happiness are the easiest to recognize [232].

Aging

Aging causes significant differences in the face such as sagging, wrinkles and spots. It is not a problem for FER studies, but it definitely creates a big issue for FR. The face texture and shape changes through the time, moreover, comes the plastic surgery to make things worse (scientifically, of course). It becomes difficult to collect data including the age factor because of the slow aging process [98], therefore although a cross-age FR is beneficial, it is yet to be implemented.

Low quality camera shooting

Although the success of face recognition systems has increased drastically, it is still difficult to analyze data coming from low quality cameras, so the shift of research must move towards surveillance-like input instead of frontal and clear face. There is also the fact that the area struggles with such extrinsic factors as well as intrinsic factors such as aging and expressions.

Deep neural networks (DNNs) have shown to perform very well on large scale object recognition problems, and face analysis has been no exception. Face analysis is a data driven task, and different face related tasks have been complementary among each other. Therefore, combining all these different tasks has been a promising research area. A DNN learns from the past, it works well on data with varying conditions, reduces complexity, and achieves higher classification rates and minimum execution time compared to other classifiers, especially on real-time or in-the-wild data where the system deals with large datasets. The only drawback of a DNN is that it consumes much more time for training than any other system, but with the help of the increased efficiency on GPU (and even TPU) computing, it became possible to train large networks with multiple layers. However, although the computer power has improved over time, there is yet another issue which is the lack of structured and balanced data. Some studies state that with the help of proper data augmentation, the lack of data and class imbalance problems could be handled well.

Analyzing the latest researches, one can easily notice that the trend of collecting and interpreting regular 2D data without any other information is being shifted to recording a 3D face shape with the depth information which may be robust to pose and illumination. It is our belief that this is the main reason the traditional single-camera-based photography is being shifted to capturing photos with 3D cameras to stream the depth information, and this will be the technological direction scientists should and will follow.

References

[1] A.H. Abdulnabi, G. Wang, J. Lu, K. Jia, Multi-task CNN model for attribute prediction, IEEE Transactions on Multimedia 17 (11) (2015) 1949–1959.

[2] Y.S. Abu-Mostafa, Learning from hints in neural networks, Journal of Complexity 6 (2) (1990) 192–198.

[3] A.S. Al-Waisy, R. Qahwaji, S. Ipson, S. Al-Fahdawi, T.A. Nagem, A multi-biometric iris recognition system based on a deep learning approach, Pattern Analysis and Applications 21 (3) (2018) 783–802.

[4] G. Antipov, M. Baccouche, J.L. Dugelay, Face aging with conditional generative adversarial networks, in: 2017 IEEE International Conference on Image Processing (ICIP), IEEE, 2017, pp. 2089–2093.

[5] A. Asthana, S. Zafeiriou, S. Cheng, M. Pantic, Incremental face alignment in the wild, in: Proceedings of the IEEE Conference on Computer Vision and Pattern Recognition, 2014, pp. 1859–1866.

[6] J. Bao, D. Chen, F. Wen, H. Li, G. Hua, Towards open-set identity preserving face synthesis, in: Proceedings of the IEEE Conference on Computer Vision and Pattern Recognition, 2018, pp. 6713–6722.

[7] S. Baron-Cohen, O. Golan, S. Wheelwright, J.J. Hill, Mind Reading: The Interactive Guide to Emotions, Jessica Kingsley, London, 2004.

[8] H. Barrett, Researching and evaluating digital storytelling as a deep learning tool, in: Society for Information Technology & Teacher Education International Conference, Association for the Advancement of Computing in Education (AACE), 2006, pp. 647–654.

[9] H. Bay, A. Ess, T. Tuytelaars, L. Van Gool, Speeded-up robust features (SURF), Computer Vision and Image Understanding 110 (3) (2008) 346–359.

[10] S. Behnke, Hierarchical Neural Networks for Image Interpretation, vol. 2766, Springer, 2003.

[11] T. Berg, P. Belhumeur, Poof: part-based one-vs.-one features for fine-grained categorization, face verification, and attribute estimation, in: Proceedings of the IEEE Conference on Computer Vision and Pattern Recognition, 2013, pp. 955–962.

[12] V. Bettadapura, Face expression recognition and analysis: the state of the art, arXiv preprint, arXiv:1203.6722, 2012.
[13] R.L. Birdwhistell, Kinesics and Context: Essays on Body Motion Communication, University of Pennsylvania Press, 1970.
[14] V. Blanz, T. Vetter, A morphable model for the synthesis of 3D faces, in: Proceedings of the 26th Annual Conference on Computer Graphics and Interactive Techniques, 1999, pp. 187–194.
[15] W.W. Bledsoe, The Model Method in Facial Recognition, Panoramic Research Inc., Palo Alto, CA, 1966, Rep. PR1, 15(47), 2.
[16] A. Boles, P. Rad, Voice biometrics: deep learning-based voiceprint authentication system, in: 2017 12th System of Systems Engineering Conference (SoSE), IEEE, 2017, pp. 1–6.
[17] L. Bourdev, J. Malik, Poselets: body part detectors trained using 3d human pose annotations, in: 2009 IEEE 12th International Conference on Computer Vision, IEEE, 2009, pp. 1365–1372.
[18] G. Bradski, A. Kaehler, OpenCV, Dr Dobb's Journal of Software Tools (2000) 3.
[19] S.C. Brubaker, J. Wu, J. Sun, M.D. Mullin, J.M. Rehg, On the design of cascades of boosted ensembles for face detection, International Journal of Computer Vision 77 (1–3) (2008) 65–86.
[20] A. Bulat, G. Tzimiropoulos, How far are we from solving the 2d & 3d face alignment problem? (and a dataset of 230,000 3d facial landmarks), in: Proceedings of the IEEE International Conference on Computer Vision, 2017, pp. 1021–1030.
[21] D. Cakir, N. Arica, Random attributes for facial expression recognition, in: 2015 International Conference on Advanced Technology & Sciences (ICAT), 2015, pp. 518–522, Proceedings of the 2nd International Conference on Advanced Technology & Sciences.
[22] D. Canedo, A.J. Neves, Facial expression recognition using computer vision: a systematic review, Applied Sciences 9 (21) (2019) 4678.
[23] J. Cao, Y. Li, Z. Zhang, Partially shared multi-task convolutional neural network with local constraint for face attribute learning, in: Proceedings of the IEEE Conference on Computer Vision and Pattern Recognition, 2018, pp. 4290–4299.
[24] Q. Cao, L. Lin, Y. Shi, X. Liang, G. Li, Attention-aware face hallucination via deep reinforcement learning, in: Proceedings of the IEEE Conference on Computer Vision and Pattern Recognition, 2017, pp. 690–698.
[25] X. Cao, Y. Wei, F. Wen, J. Sun, Face alignment by explicit shape regression, International Journal of Computer Vision 107 (2) (2014) 177–190.
[26] N. Caporusso, K. Zhang, G. Carlson, D. Jachetta, D. Patchin, S. Romeiser, N. Vaughn, A. Walters, User discrimination of content produced by generative adversarial networks, in: International Conference on Human Interaction and Emerging Technologies, Springer, Cham, 2019, pp. 725–730.
[27] A. Celik, N. Arica, Enhancing face pose normalization with deep learning, Turkish Journal of Electrical Engineering & Computer Sciences 27 (5) (2019) 3699–3712.
[28] F.J. Chang, A.T. Tran, T. Hassner, I. Masi, R. Nevatia, G. Medioni, Deep, landmark-free fame: face alignment, modeling, and expression estimation, International Journal of Computer Vision 127 (6–7) (2019) 930–956.
[29] A. Chaudhuri, Deep learning models for face recognition: a comparative analysis, in: Deep Biometrics, Springer, Cham, 2020, pp. 99–140.
[30] D. Chen, X. Cao, L. Wang, F. Wen, J. Sun, Bayesian face revisited: a joint formulation, in: European Conference on Computer Vision, Springer, Berlin, Heidelberg, 2012, pp. 566–579.
[31] J. Chen, X. Kang, Y. Liu, Z.J. Wang, Median filtering forensics based on convolutional neural networks, IEEE Signal Processing Letters 22 (11) (2015) 1849–1853.

[32] Y. Chen, Y. Tai, X. Liu, C. Shen, J. Yang, Fsrnet: end-to-end learning face super-resolution with facial priors, in: Proceedings of the IEEE Conference on Computer Vision and Pattern Recognition, 2018, pp. 2492–2501.
[33] E. Chong, N. Ruiz, Y. Wang, Y. Zhang, A. Rozga, J.M. Rehg, Connecting gaze, scene, and attention: generalized attention estimation via joint modeling of gaze and scene saliency, in: Proceedings of the European Conference on Computer Vision (ECCV), 2018, pp. 383–398.
[34] W.S. Chu, F. De la Torre, J.F. Cohn, Selective transfer machine for personalized facial action unit detection, in: Proceedings of the IEEE Conference on Computer Vision and Pattern Recognition, 2013, pp. 3515–3522.
[35] D.C. Ciresan, U. Meier, J. Masci, L.M. Gambardella, J. Schmidhuber, Flexible, high performance convolutional neural networks for image classification, in: Twenty-Second International Joint Conference on Artificial Intelligence, 2011.
[36] T.F. Cootes, G.J. Edwards, C.J. Taylor, Active appearance models, IEEE Transactions on Pattern Analysis and Machine Intelligence 23 (6) (2001) 681–685.
[37] T.F. Cootes, M.C. Ionita, C. Lindner, P. Sauer, Robust and accurate shape model fitting using random forest regression voting, in: European Conference on Computer Vision, Springer, Berlin, Heidelberg, 2012, pp. 278–291.
[38] C.A. Corneanu, M.O. Simón, J.F. Cohn, S.E. Guerrero, Survey on rgb, 3d, thermal, and multimodal approaches for facial expression recognition: history, trends, and affect-related applications, IEEE Transactions on Pattern Analysis and Machine Intelligence 38 (8) (2016) 1548–1568.
[39] Charles Darwin, The Expression of the Emotions in Man and Animals, Murray, London, UK, 1872, 3rd ed. P. Ekman, London, UK: HarperCollins, 1998.
[40] R. Das, A. Gadre, S. Zhang, S. Kumar, J.M. Moura, A deep learning approach to IoT authentication, in: 2018 IEEE International Conference on Communications (ICC), IEEE, 2018, pp. 1–6.
[41] J. Deng, S. Cheng, N. Xue, Y. Zhou, S. Zafeiriou, Uv-gan: adversarial facial uv map completion for pose-invariant face recognition, in: Proceedings of the IEEE Conference on Computer Vision and Pattern Recognition, 2018, pp. 7093–7102.
[42] J. Deng, Y. Zhou, S. Zafeiriou, Marginal loss for deep face recognition, in: Proceedings of the IEEE Conference on Computer Vision and Pattern Recognition Workshops, 2017, pp. 60–68.
[43] A. Dhall, R. Goecke, S. Ghosh, J. Joshi, J. Hoey, T. Gedeon, From individual to group-level emotion recognition: Emotiw 5.0, in: Proceedings of the 19th ACM International Conference on Multimodal Interaction, 2017, pp. 524–528.
[44] H. Dibeklioglu, A. Ali Salah, T. Gevers, Like father, like son: facial expression dynamics for kinship verification, in: Proceedings of the IEEE International Conference on Computer Vision, 2013, pp. 1497–1504.
[45] C. Ding, D. Tao, A comprehensive survey on pose-invariant face recognition, ACM Transactions on Intelligent Systems and Technology (TIST) 7 (3) (2016) 1–42.
[46] H. Ding, H. Zhou, S.K. Zhou, R. Chellappa, A deep cascade network for unaligned face attribute classification, in: Thirty-Second AAAI Conference on Artificial Intelligence, 2018.
[47] H. Ding, S.K. Zhou, R. Chellappa, Facenet2expnet: regularizing a deep face recognition net for expression recognition, in: 2017 12th IEEE International Conference on Automatic Face & Gesture Recognition (FG 2017), IEEE, 2017, pp. 118–126.
[48] G. Donato, M.S. Bartlett, J.C. Hager, P. Ekman, T.J. Sejnowski, Classifying facial actions, IEEE Transactions on Pattern Analysis and Machine Intelligence 21 (10) (1999) 974–989.
[49] P. Dou, I.A. Kakadiaris, Multi-view 3D face reconstruction with deep recurrent neural networks, Image and Vision Computing 80 (2018) 80–91.

[50] K. Dwivedi, K. Biswaranjan, A. Sethi, Drowsy driver detection using representation learning, in: 2014 IEEE International Advance Computing Conference (IACC), IEEE, 2014, pp. 995–999.
[51] S. Ebrahimi Kahou, V. Michalski, K. Konda, R. Memisevic, C. Pal, Recurrent neural networks for emotion recognition in video, in: Proceedings of the 2015 ACM on International Conference on Multimodal Interaction, 2015, pp. 467–474.
[52] M. Ehrlich, T.J. Shields, T. Almaev, M.R. Amer, Facial attributes classification using multi-task representation learning, in: Proceedings of the IEEE Conference on Computer Vision and Pattern Recognition Workshops, 2016, pp. 47–55.
[53] Paul Ekman, Wallace V. Friesen, Unmasking the Face: A Guide to Recognizing Emotions From Facial Cues, 1975.
[54] G. Fanelli, T. Weise, J. Gall, L. Van Gool, Real time head pose estimation from consumer depth cameras, in: Joint Pattern Recognition Symposium, Springer, Berlin, Heidelberg, 2011, pp. 101–110.
[55] S.S. Farfade, M.J. Saberian, L.J. Li, Multi-view face detection using deep convolutional neural networks, in: Proceedings of the 5th ACM on International Conference on Multimedia Retrieval, 2015, pp. 643–650.
[56] A. Farhadi, I. Endres, D. Hoiem, D. Forsyth, Describing objects by their attributes, in: 2009 IEEE Conference on Computer Vision and Pattern Recognition, IEEE, 2009, pp. 1778–1785.
[57] P.F. Felzenszwalb, R.B. Girshick, D. McAllester, D. Ramanan, Object detection with discriminatively trained part-based models, IEEE Transactions on Pattern Analysis and Machine Intelligence 32 (9) (2009) 1627–1645.
[58] A. Ferdowsi, W. Saad, Deep learning-based dynamic watermarking for secure signal authentication in the Internet of Things, in: 2018 IEEE International Conference on Communications (ICC), IEEE, 2018, pp. 1–6.
[59] Y. Freund, R.E. Schapire, A decision-theoretic generalization of on-line learning and an application to boosting, in: European Conference on Computational Learning Theory, Springer, Berlin, Heidelberg, 1995, pp. 23–37.
[60] L. Fridman, D.E. Brown, M. Glazer, W. Angell, S. Dodd, B. Jenik, et al., MIT autonomous vehicle technology study: large-scale deep learning based analysis of driver behavior and interaction with automation, arXiv preprint, arXiv:1711.06976, 2017, 1.
[61] P. Ekman, E. Friesen, Facial action coding system: a technique for the measurement of facial movement, Palo Alto, 1978, 3.
[62] K. Fukushima, Neocognitron: a self-organizing neural network model for a mechanism of pattern recognition unaffected by shift in position, Biological Cybernetics 36 (4) (1980) 193–202.
[63] C. Galea, R.A. Farrugia, Forensic face photo-sketch recognition using a deep learning-based architecture, IEEE Signal Processing Letters 24 (11) (2017) 1586–1590.
[64] C. Garcia, M. Delakis, Convolutional face finder: a neural architecture for fast and robust face detection, IEEE Transactions on Pattern Analysis and Machine Intelligence 26 (11) (2004) 1408–1423.
[65] F.A. Gers, J. Schmidhuber, F. Cummins, Learning to forget: continual prediction with LSTM, 1999.
[66] R. Girshick, Fast r-cnn, in: Proceedings of the IEEE International Conference on Computer Vision, 2015, pp. 1440–1448.
[67] R. Girshick, J. Donahue, T. Darrell, J. Malik, Rich feature hierarchies for accurate object detection and semantic segmentation, in: Proceedings of the IEEE Conference on Computer Vision and Pattern Recognition, 2014, pp. 580–587.

[68] R. Girshick, F. Iandola, T. Darrell, J. Malik, Deformable part models are convolutional neural networks, in: Proceedings of the IEEE Conference on Computer Vision and Pattern Recognition, 2015, pp. 437–446.
[69] I. Goodfellow, J. Pouget-Abadie, M. Mirza, B. Xu, D. Warde-Farley, S. Ozair, et al., Generative adversarial nets, in: Advances in Neural Information Processing Systems, 2014, pp. 2672–2680.
[70] C. Gordon, R. Debus, Developing deep learning approaches and personal teaching efficacy within a preservice teacher education context, British Journal of Educational Psychology 72 (4) (2002) 483–511.
[71] M. Gori, G. Monfardini, F. Scarselli, A new model for learning in graph domains, in: 2005 IEEE International Joint Conference on Neural Networks, Proceedings, vol. 2, IEEE, 2005, pp. 729–734.
[72] A. Graves, J. Schmidhuber, Offline handwriting recognition with multidimensional recurrent neural networks, in: Advances in Neural Information Processing Systems, 2009, pp. 545–552.
[73] H. Greenspan, B. Van Ginneken, R.M. Summers, Guest editorial deep learning in medical imaging: overview and future promise of an exciting new technique, IEEE Transactions on Medical Imaging 35 (5) (2016) 1153–1159.
[74] M. Günther, A. Rozsa, T.E. Boult, AFFACT: alignment-free facial attribute classification technique, in: 2017 IEEE International Joint Conference on Biometrics (IJCB), IEEE, 2017, pp. 90–99.
[75] W. Guo, D. Mu, J. Xu, P. Su, G. Wang, X. Xing, Lemna: explaining deep learning based security applications, in: Proceedings of the 2018 ACM SIGSAC Conference on Computer and Communications Security, 2018, pp. 364–379.
[76] H. Han, A.K. Jain, F. Wang, S. Shan, X. Chen, Heterogeneous face attribute estimation: a deep multi-task learning approach, IEEE Transactions on Pattern Analysis and Machine Intelligence 40 (11) (2017) 2597–2609.
[77] B. Hasani, M.H. Mahoor, Facial expression recognition using enhanced deep 3D convolutional neural networks, in: Proceedings of the IEEE Conference on Computer Vision and Pattern Recognition Workshops, 2017, pp. 30–40.
[78] M. Hasnat, J. Bohné, J. Milgram, S. Gentric, L. Chen, von Mises-Fisher mixture model-based deep learning: application to face verification, arXiv preprint, arXiv: 1706.04264, 2017.
[79] T. Hassner, S. Harel, E. Paz, R. Enbar, Effective face frontalization in unconstrained images, in: Proceedings of the IEEE Conference on Computer Vision and Pattern Recognition, 2015, pp. 4295–4304.
[80] J. He, D. Li, B. Yang, S. Cao, B. Sun, L. Yu, Multi view facial action unit detection based on CNN and BLSTM-RNN, in: 2017 12th IEEE International Conference on Automatic Face & Gesture Recognition (FG 2017), IEEE, 2017, pp. 848–853.
[81] K. He, Y. Fu, W. Zhang, C. Wang, Y.G. Jiang, F. Huang, X. Xue, Harnessing synthesized abstraction images to improve facial attribute recognition, in: IJCAI, 2018, pp. 733–740.
[82] G.E. Hinton, S. Osindero, Y.W. Teh, A fast learning algorithm for deep belief nets, Neural Computation 18 (7) (2006) 1527–1554.
[83] S. Hochreiter, J. Schmidhuber, Long short-term memory, Neural Computation 9 (8) (1997) 1735–1780.
[84] R. Huang, S. Zhang, T. Li, R. He, Beyond face rotation: global and local perception gan for photorealistic and identity preserving frontal view synthesis, in: Proceedings of the IEEE International Conference on Computer Vision, 2017, pp. 2439–2448.
[85] E.P. Ijjina, C.K. Mohan, Facial expression recognition using kinect depth sensor and convolutional neural networks, in: 2014 13th International Conference on Machine Learning and Applications, IEEE, 2014, pp. 392–396.

[86] R.E. Jack, O.G. Garrod, H. Yu, R. Caldara, P.G. Schyns, Facial expressions of emotion are not culturally universal, Proceedings of the National Academy of Sciences 109 (19) (2012) 7241–7244.
[87] A. Jourabloo, X. Liu, Large-pose face alignment via CNN-based dense 3D model fitting, in: Proceedings of the IEEE Conference on Computer Vision and Pattern Recognition, 2016, pp. 4188–4196.
[88] H. Jung, S. Lee, J. Yim, S. Park, J. Kim, Joint fine-tuning in deep neural networks for facial expression recognition, in: Proceedings of the IEEE International Conference on Computer Vision, 2015, pp. 2983–2991.
[89] M.M. Kalayeh, B. Gong, M. Shah, Improving facial attribute prediction using semantic segmentation, in: Proceedings of the IEEE Conference on Computer Vision and Pattern Recognition, 2017, pp. 6942–6950.
[90] M.J. Kang, J.W. Kang, Intrusion detection system using deep neural network for in-vehicle network security, PLoS ONE 11 (6) (2016).
[91] V. Kazemi, J. Sullivan, One millisecond face alignment with an ensemble of regression trees, in: Proceedings of the IEEE Conference on Computer Vision and Pattern Recognition, 2014, pp. 1867–1874.
[92] D.E. King, Dlib-ml: a machine learning toolkit, Journal of Machine Learning Research 10 (2009) 1755–1758.
[93] A. Krizhevsky, I. Sutskever, G.E. Hinton, Imagenet classification with deep convolutional neural networks, in: Advances in Neural Information Processing Systems, 2012, pp. 1097–1105.
[94] A. Kumar, A. Alavi, R. Chellappa, Kepler: keypoint and pose estimation of unconstrained faces by learning efficient h-cnn regressors, in: 2017 12th IEEE International Conference on Automatic Face & Gesture Recognition (FG 2017), IEEE, 2017, pp. 258–265.
[95] N. Kumar, A.C. Berg, P.N. Belhumeur, S.K. Nayar, Attribute and smile classifiers for face verification, in: 2009 IEEE 12th International Conference on Computer Vision, IEEE, 2009, pp. 365–372.
[96] H. Lai, S. Xiao, Z. Cui, Y. Pan, C. Xu, S. Yan, Deep cascaded regression for face alignment, arXiv preprint, arXiv:1510.09083, 2015, 1.
[97] N.D. Lane, S. Bhattacharya, P. Georgiev, C. Forlivesi, L. Jiao, L. Qendro, F. Kawsar, Deepx: a software accelerator for low-power deep learning inference on mobile devices, in: 2016 15th ACM/IEEE International Conference on Information Processing in Sensor Networks (IPSN), IEEE, 2016, pp. 1–12.
[98] A. Lanitis, C.J. Taylor, T.F. Cootes, Toward automatic simulation of aging effects on face images, IEEE Transactions on Pattern Analysis and Machine Intelligence 24 (4) (2002) 442–455.
[99] S. Lawrence, C.L. Giles, A.C. Tsoi, A.D. Back, Face recognition: a convolutional neural-network approach, IEEE Transactions on Neural Networks 8 (1) (1997) 98–113.
[100] Y. LeCun, B. Boser, J.S. Denker, D. Henderson, R.E. Howard, W. Hubbard, L.D. Jackel, Backpropagation applied to handwritten zip code recognition, Neural Computation 1 (4) (1989) 541–551.
[101] Y. LeCun, L. Bottou, Y. Bengio, P. Haffner, Gradient-based learning applied to document recognition, Proceedings of the IEEE 86 (11) (1998) 2278–2324.
[102] J.J. Lee, M.Z. Uddin, T.S. Kim, Spatiotemporal human facial expression recognition using Fisher independent component analysis and hidden Markov model, in: 2008 30th Annual International Conference of the IEEE Engineering in Medicine and Biology Society, IEEE, 2008, pp. 2546–2549.
[103] I. Lenz, H. Lee, A. Saxena, Deep learning for detecting robotic grasps, The International Journal of Robotics Research 34 (4–5) (2015) 705–724.

[104] A.L. Levada, D.C. Correa, D.H. Salvadeo, J.H. Saito, N.D. Mascarenhas, Novel approaches for face recognition: template-matching using dynamic time warping and LSTM Neural Network Supervised Classification, in: 2008 15th International Conference on Systems, Signals and Image Processing, IEEE, 2008, pp. 241–244.
[105] G. Levi, T. Hassner, Emotion recognition in the wild via convolutional neural networks and mapped binary patterns, in: Proceedings of the 2015 ACM on International Conference on Multimodal Interaction, 2015, pp. 503–510.
[106] S. Levine, P. Pastor, A. Krizhevsky, J. Ibarz, D. Quillen, Learning hand-eye coordination for robotic grasping with deep learning and large-scale data collection, The International Journal of Robotics Research 37 (4–5) (2018) 421–436.
[107] D. Li, X. Chen, K. Huang, Multi-attribute learning for pedestrian attribute recognition in surveillance scenarios, in: 2015 3rd IAPR Asian Conference on Pattern Recognition (ACPR), IEEE, 2015, pp. 111–115.
[108] H. Li, Z. Lin, X. Shen, J. Brandt, G. Hua, A convolutional neural network cascade for face detection, in: Proceedings of the IEEE Conference on Computer Vision and Pattern Recognition, 2015, pp. 5325–5334.
[109] P. Li, J. Li, Z. Huang, T. Li, C.Z. Gao, S.M. Yiu, K. Chen, Multi-key privacy-preserving deep learning in cloud computing, Future Generations Computer Systems 74 (2017) 76–85.
[110] S. Li, W. Deng, Deep facial expression recognition: a survey, arXiv preprint, arXiv:1804.08348, 2018.
[111] W. Li, M. Li, Z. Su, Z. Zhu, A deep-learning approach to facial expression recognition with candid images, in: 2015 14th IAPR International Conference on Machine Vision Applications (MVA), IEEE, 2015, pp. 279–282.
[112] L. Lin, D. Zhang, P. Luo, W. Zuo, Face localization and enhancement, in: Human Centric Visual Analysis With Deep Learning, Springer, Singapore, 2020, pp. 29–45.
[113] G. Litjens, T. Kooi, B.E. Bejnordi, A.A.A. Setio, F. Ciompi, M. Ghafoorian, C.I. Sánchez, A survey on deep learning in medical image analysis, Medical Image Analysis 42 (2017) 60–88.
[114] C. Liu, D. Sun, On Bayesian adaptive video super resolution, IEEE Transactions on Pattern Analysis and Machine Intelligence 36 (2) (2013) 346–360.
[115] J. Liu, Y. Deng, T. Bai, Z. Wei, C. Huang, Targeting ultimate accuracy: face recognition via deep embedding, arXiv preprint, arXiv:1506.07310, 2015.
[116] K. Liu, M. Zhang, Z. Pan, Facial expression recognition with CNN ensemble, in: 2016 International Conference on Cyberworlds (CW), IEEE, 2016, pp. 163–166.
[117] M. Liu, S. Li, S. Shan, X. Chen, Au-inspired deep networks for facial expression feature learning, Neurocomputing 159 (2015) 126–136.
[118] P. Liu, S. Han, Z. Meng, Y. Tong, Facial expression recognition via a boosted deep belief network, in: Proceedings of the IEEE Conference on Computer Vision and Pattern Recognition, 2014, pp. 1805–1812.
[119] W. Liu, D. Anguelov, D. Erhan, C. Szegedy, S. Reed, C.Y. Fu, A.C. Berg, Ssd: Single shot multibox detector, in: European Conference on Computer Vision, Springer, Cham, 2016, pp. 21–37.
[120] W. Liu, C. Song, Y. Wang, Facial expression recognition based on discriminative dictionary learning, in: Proceedings of the 21st International Conference on Pattern Recognition (ICPR2012), IEEE, 2012, pp. 1839–1842.
[121] W. Liu, Y. Wen, Z. Yu, M. Li, B. Raj, L. Song, Sphereface: deep hypersphere embedding for face recognition, in: Proceedings of the IEEE Conference on Computer Vision and Pattern Recognition, 2017, pp. 212–220.
[122] X. Liu, W. Liu, T. Mei, H. Ma, A deep learning-based approach to progressive vehicle re-identification for urban surveillance, in: European Conference on Computer Vision, Springer, Cham, 2016, pp. 869–884.

[123] Y. Liu, H. Li, X. Wang, Rethinking feature discrimination and polymerization for large-scale recognition, arXiv preprint, arXiv:1710.00870, 2017.
[124] Y. Liu, Q. Li, Z. Sun, Attribute-aware face aging with wavelet-based generative adversarial networks, in: Proceedings of the IEEE Conference on Computer Vision and Pattern Recognition, 2019, pp. 11877–11886.
[125] Y. Liu, J. Ling, Z. Liu, J. Shen, C. Gao, Finger vein secure biometric template generation based on deep learning, Soft Computing 22 (7) (2018) 2257–2265.
[126] Z. Liu, P. Luo, X. Wang, X. Tang, Deep learning face attributes in the wild, in: Proceedings of the IEEE International Conference on Computer Vision, 2015, pp. 3730–3738.
[127] A.T. Lopes, E. de Aguiar, A.F. De Souza, T. Oliveira-Santos, Facial expression recognition with convolutional neural networks: coping with few data and the training sample order, Pattern Recognition 61 (2017) 610–628.
[128] D.G. Lowe, Distinctive image features from scale-invariant keypoints, International Journal of Computer Vision 60 (2) (2004) 91–110.
[129] Y. Lu, A. Kumar, S. Zhai, Y. Cheng, T. Javidi, R. Feris, Fully-adaptive feature sharing in multi-task networks with applications in person attribute classification, in: Proceedings of the IEEE Conference on Computer Vision and Pattern Recognition, 2017, pp. 5334–5343.
[130] P. Lucey, J.F. Cohn, T. Kanade, J. Saragih, Z. Ambadar, I. Matthews, The extended Cohn–Kanade dataset (ck+): a complete dataset for action unit and emotion-specified expression, in: 2010 IEEE Computer Society Conference on Computer Vision and Pattern Recognition-Workshops, IEEE, 2010, pp. 94–101.
[131] P. Luo, X. Wang, X. Tang, A deep sum-product architecture for robust facial attributes analysis, in: Proceedings of the IEEE International Conference on Computer Vision, 2013, pp. 2864–2871.
[132] Y. Lv, Z. Feng, C. Xu, Facial expression recognition via deep learning, in: 2014 International Conference on Smart Computing, IEEE, 2014, pp. 303–308.
[133] U. Mahbub, S. Sarkar, R. Chellappa, Segment-based methods for facial attribute detection from partial faces, IEEE Transactions on Affective Computing (2018).
[134] F. Mamalet, S. Roux, C. Garcia, Embedded facial image processing with convolutional neural networks, in: Proceedings of 2010 IEEE International Symposium on Circuits and Systems, IEEE, 2010, pp. 261–264.
[135] M. Martin, F. Van De Camp, R. Stiefelhagen, Real time head model creation and head pose estimation on consumer depth cameras, in: 2014 2nd International Conference on 3D Vision, Vol. 1, IEEE, 2014, pp. 641–648.
[136] I. Masi, Y. Wu, T. Hassner, P. Natarajan, Deep face recognition: a survey, in: 2018 31st SIBGRAPI Conference on Graphics, Patterns and Images (SIBGRAPI), IEEE, 2018, pp. 471–478.
[137] M. Matsugu, K. Mori, Y. Mitari, Y. Kaneda, Subject independent facial expression recognition with robust face detection using a convolutional neural network, Neural Networks 16 (5–6) (2003) 555–559.
[138] V. Mavani, S. Raman, K.P. Miyapuram, Facial expression recognition using visual saliency and deep learning, in: Proceedings of the IEEE International Conference on Computer Vision Workshops, 2017, pp. 2783–2788.
[139] A. Mehrabian, Communication without words, Communication Theory (1968) 193–200.
[140] Z. Meng, P. Liu, J. Cai, S. Han, Y. Tong, Identity-aware convolutional neural network for facial expression recognition, in: 2017 12th IEEE International Conference on Automatic Face & Gesture Recognition (FG 2017), IEEE, 2017, pp. 558–565.
[141] G.P. Meyer, S. Gupta, I. Frosio, D. Reddy, J. Kautz, Robust model-based 3d head pose estimation, in: Proceedings of the IEEE International Conference on Computer Vision, 2015, pp. 3649–3657.

[142] A. Mollahosseini, D. Chan, M.H. Mahoor, Going deeper in facial expression recognition using deep neural networks, in: 2016 IEEE Winter Conference on Applications of Computer Vision (WACV), IEEE, 2016, pp. 1–10.
[143] S.S. Mukherjee, N.M. Robertson, Deep head pose: gaze-direction estimation in multimodal video, IEEE Transactions on Multimedia 17 (11) (2015) 2094–2107.
[144] M. Murtaza, M. Sharif, M. Raza, J.H. Shah, Analysis of face recognition under varying facial expression: a survey, The International Arab Journal of Information Technology 10 (4) (2013) 378–388.
[145] R. Natsume, T. Yatagawa, S. Morishima, RSGAN: face swapping and editing using face and hair representation in latent spaces, arXiv preprint, arXiv:1804.03447, 2018.
[146] N. Neverova, C. Wolf, G.W. Taylor, F. Nebout, Multi-scale deep learning for gesture detection and localization, in: European Conference on Computer Vision, Springer, Cham, 2014, pp. 474–490.
[147] C. Nhan Duong, K. Luu, K. Gia Quach, T.D. Bui, Beyond principal components: deep Boltzmann machines for face modeling, in: Proceedings of the IEEE Conference on Computer Vision and Pattern Recognition, 2015, pp. 4786–4794.
[148] X. Niu, H. Han, S. Shan, X. Chen, Multi-label co-regularization for semi-supervised facial action unit recognition, in: Advances in Neural Information Processing Systems, 2019, pp. 907–917.
[149] A. Ortony, T.J. Turner, What's basic about basic emotions?, Psychological Review 97 (3) (1990) 315.
[150] M. Padmanabhan, Intraface: negotiating gender-relations in agrobiodiversity, FZG – Freiburger Zeitschrift für GeschlechterStudien 22 (2) (2016).
[151] J. Park, J.Y. Lee, D. Yoo, I. So Kweon, Distort-and-recover: color enhancement using deep reinforcement learning, in: Proceedings of the IEEE Conference on Computer Vision and Pattern Recognition, 2018, pp. 5928–5936.
[152] O.M. Parkhi, A. Vedaldi, A. Zisserman, Deep face recognition, 2015.
[153] W.G. Parrott (Ed.), Emotions in Social Psychology: Essential Readings, Psychology Press, 2001.
[154] M. Patacchiola, A. Cangelosi, Head pose estimation in the wild using convolutional neural networks and adaptive gradient methods, Pattern Recognition 71 (2017) 132–143.
[155] M.T. Pham, T.J. Cham, Fast training and selection of Haar features using statistics in boosting-based face detection, in: 2007 IEEE 11th International Conference on Computer Vision, IEEE, 2007, pp. 1–7.
[156] I. Kotsia, I. Buciu, I. Pitas, An analysis of facial expression recognition under partial facial image occlusion, in: 2008 Image and Vision Computing, 2008, pp. 1052–1067.
[157] G. Pons, D. Masip, Multi-task, multi-label and multi-domain learning with residual convolutional networks for emotion recognition, arXiv preprint, arXiv:1802.06664, 2018.
[158] B. Prihasto, S. Choirunnisa, M.I. Nurdiansyah, S. Mathulaprangsan, V.C.M. Chu, S.H. Chen, J.C. Wang, A survey of deep face recognition in the wild, in: 2016 International Conference on Orange Technologies (ICOT), IEEE, 2016, pp. 76–79.
[159] A. Punitha, M.K. Geetha, HMM based real time facial expression recognition, International Journal of Emerging Technology and Advanced Engineering 3 (1) (2013) 180–185.
[160] X. Qi, L. Zhang, Face recognition via centralized coordinate learning, arXiv preprint, arXiv:1801.05678, 2018.
[161] S. Rajan, P. Chenniappan, S. Devaraj, N. Madian, Facial expression recognition techniques: a comprehensive survey, IET Image Processing 13 (7) (2019) 1031–1040.
[162] X. Ran, H. Chen, X. Zhu, Z. Liu, J. Chen, Deepdecision: a mobile deep learning framework for edge video analytics, in: IEEE INFOCOM 2018-IEEE Conference on Computer Communications, IEEE, 2018, pp. 1421–1429.

[163] R. Ranjan, C.D. Castillo, R. Chellappa, L2-constrained softmax loss for discriminative face verification, arXiv preprint, arXiv:1703.09507, 2017.

[164] R. Ranjan, V.M. Patel, R. Chellappa, A deep pyramid deformable part model for face detection, in: 2015 IEEE 7th International Conference on Biometrics Theory, Applications and Systems (BTAS), IEEE, 2015, pp. 1–8.

[165] R. Ranjan, V.M. Patel, R. Chellappa, Hyperface: a deep multi-task learning framework for face detection, landmark localization, pose estimation, and gender recognition, IEEE Transactions on Pattern Analysis and Machine Intelligence 41 (1) (2017) 121–135.

[166] R. Ranjan, S. Sankaranarayanan, A. Bansal, N. Bodla, J.C. Chen, V.M. Patel, R. Chellappa, Deep learning for understanding faces: machines may be just as good, or better, than humans, IEEE Signal Processing Magazine 35 (1) (2018) 66–83.

[167] R. Ranjan, S. Sankaranarayanan, C.D. Castillo, R. Chellappa, An all-in-one convolutional neural network for face analysis, in: 2017 12th IEEE International Conference on Automatic Face & Gesture Recognition (FG 2017), IEEE, 2017, pp. 17–24.

[168] Y. Rao, J. Lu, J. Zhou, Attention-aware deep reinforcement learning for video face recognition, in: Proceedings of the IEEE International Conference on Computer Vision, 2017, pp. 3931–3940.

[169] D. Ravì, C. Wong, F. Deligianni, M. Berthelot, J. Andreu-Perez, B. Lo, G.Z. Yang, Deep learning for health informatics, IEEE Journal of Biomedical and Health Informatics 21 (1) (2016) 4–21.

[170] B. Reddy, Y.H. Kim, S. Yun, C. Seo, J. Jang, Real-time driver drowsiness detection for embedded system using model compression of deep neural networks, in: Proceedings of the IEEE Conference on Computer Vision and Pattern Recognition Workshops, 2017, pp. 121–128.

[171] J. Ren, Y. Hu, Y.W. Tai, C. Wang, L. Xu, W. Sun, Q. Yan, Look, listen and learn—a multimodal LSTM for speaker identification, in: Thirtieth AAAI Conference on Artificial Intelligence, 2016.

[172] S. Ren, K. He, R. Girshick, J. Sun, Faster r-cnn: towards real-time object detection with region proposal networks, in: Advances in Neural Information Processing Systems, 2015, pp. 91–99.

[173] B.D. Rouhani, M.S. Riazi, F. Koushanfar, Deepsecure: scalable provably-secure deep learning, in: Proceedings of the 55th Annual Design Automation Conference, 2018, pp. 1–6.

[174] E.M. Rudd, M. Günther, T.E. Boult, Moon: a mixed objective optimization network for the recognition of facial attributes, in: European Conference on Computer Vision, Springer, Cham, 2016, pp. 19–35.

[175] S. Ruder, An overview of multi-task learning in deep neural networks, arXiv preprint, arXiv:1706.05098, 2017.

[176] D.E. Rumelhart, G.E. Hinton, R.J. Williams, Learning representations by back-propagating errors, Nature 323 (6088) (1986) 533–536.

[177] F. Schroff, D. Kalenichenko, J. Philbin, Facenet: a unified embedding for face recognition and clustering, in: Proceedings of the IEEE Conference on Computer Vision and Pattern Recognition, 2015, pp. 815–823.

[178] N. Sebe, I. Cohen, T.S. Huang, Multimodal emotion recognition, in: Handbook of Pattern Recognition and Computer Vision, 2005, pp. 387–409.

[179] C. Shan, S. Gong, P.W. McOwan, Facial expression recognition based on local binary patterns: a comprehensive study, Image and Vision Computing 27 (6) (2009) 803–816.

[180] P.Y. Simard, D. Steinkraus, J.C. Platt, Best practices for convolutional neural networks applied to visual document analysis, in: Icdar, Vol. 3, No. 2003, 2003.

[181] T. Simon, M.H. Nguyen, F. De La Torre, J.F. Cohn, Action unit detection with segment-based svms, in: 2010 IEEE Computer Society Conference on Computer Vision and Pattern Recognition, IEEE, 2010, pp. 2737–2744.
[182] C.P. Sumathi, T. Santhanam, M. Mahadevi, Automatic facial expression analysis a survey, International Journal of Computer Science and Engineering Survey 3 (6) (2012) 47.
[183] B. Sun, Q. Wei, L. Li, Q. Xu, J. He, L. Yu, LSTM for dynamic emotion and group emotion recognition in the wild, in: Proceedings of the 18th ACM International Conference on Multimodal Interaction, 2016, pp. 451–457.
[184] Y. Sun, Y. Chen, X. Wang, X. Tang, Deep learning face representation by joint identification-verification, in: Advances in Neural Information Processing Systems, 2014, pp. 1988–1996.
[185] Y. Sun, D. Liang, X. Wang, X. Tang, Deepid3: face recognition with very deep neural networks, arXiv preprint, arXiv:1502.00873, 2015.
[186] Y. Sun, X. Wang, X. Tang, Deep convolutional network cascade for facial point detection, in: Proceedings of the IEEE Conference on Computer Vision and Pattern Recognition, 2013, pp. 3476–3483.
[187] L. Surace, M. Patacchiola, E. Battini Sönmez, W. Spataro, A. Cangelosi, Emotion recognition in the wild using deep neural networks and Bayesian classifiers, in: Proceedings of the 19th ACM International Conference on Multimodal Interaction, 2017, pp. 593–597.
[188] C. Szegedy, W. Liu, Y. Jia, P. Sermanet, S. Reed, D. Anguelov, A. Rabinovich, Going deeper with convolutions, in: Proceedings of the IEEE Conference on Computer Vision and Pattern Recognition, 2015, pp. 1–9.
[189] Y. Taigman, M. Yang, M.A. Ranzato, L. Wolf, Deepface: closing the gap to human-level performance in face verification, in: Proceedings of the IEEE Conference on Computer Vision and Pattern Recognition, 2014.
[190] G. Tesauro, Temporal difference learning and TD-Gammon, Communications of the ACM 38 (3) (1995) 58–68.
[191] Y. Tian, X. Peng, L. Zhao, S. Zhang, D.N. Metaxas, CR-GAN: learning complete representations for multi-view generation, arXiv preprint, arXiv:1806.11191, 2018.
[192] L. Tran, X. Yin, X. Liu, Disentangled representation learning gan for pose-invariant face recognition, in: Proceedings of the IEEE Conference on Computer Vision and Pattern Recognition, 2017, pp. 1415–1424.
[193] G. Trigeorgis, P. Snape, M.A. Nicolaou, E. Antonakos, S. Zafeiriou, Mnemonic descent method: a recurrent process applied for end-to-end face alignment, in: Proceedings of the IEEE Conference on Computer Vision and Pattern Recognition, 2016, pp. 4177–4187.
[194] P.C. Trimmer, E.S. Paul, M.T. Mendl, J.M. McNamara, A.I. Houston, On the evolution and optimality of mood states, Behavioral Sciences 3 (3) (2013) 501–521.
[195] M. Turk, A. Pentland, Eigenfaces for recognition, Journal of Cognitive Neuroscience 3 (1) (1991) 71–86.
[196] J.R. Uijlings, K.E. Van De Sande, T. Gevers, A.W. Smeulders, Selective search for object recognition, International Journal of Computer Vision 104 (2) (2013) 154–171.
[197] B. Van Ginneken, A.F. Frangi, J.J. Staal, B.M. ter Haar Romeny, M.A. Viergever, Active shape model segmentation with optimal features, IEEE Transactions on Medical Imaging 21 (8) (2002) 924–933.
[198] O. Vinyals, A. Toshev, S. Bengio, D. Erhan, Show and tell: a neural image caption generator, in: Proceedings of the IEEE Conference on Computer Vision and Pattern Recognition, 2015, pp. 3156–3164.
[199] P. Viola, M.J. Jones, Robust real-time face detection, International Journal of Computer Vision 57 (2) (2004) 137–154.

[200] A.S. Vyas, H.B. Prajapati, V.K. Dabhi, Survey on face expression recognition using CNN, in: 2019 5th International Conference on Advanced Computing & Communication Systems (ICACCS), IEEE, 2019, pp. 102–106.

[201] F. Wang, J. Cheng, W. Liu, H. Liu, Additive margin softmax for face verification, IEEE Signal Processing Letters 25 (7) (2018) 926–930.

[202] F. Wang, X. Xiang, J. Cheng, A.L. Yuille, Normface: L2 hypersphere embedding for face verification, in: Proceedings of the 25th ACM International Conference on Multimedia, 2017, pp. 1041–1049.

[203] H. Wang, Y. Wang, Z. Zhou, X. Ji, D. Gong, J. Zhou, W. Liu, Cosface: large margin cosine loss for deep face recognition, in: Proceedings of the IEEE Conference on Computer Vision and Pattern Recognition, 2018, pp. 5265–5274.

[204] M. Wang, W. Deng, Deep face recognition: a survey, arXiv preprint, arXiv:1804.06655, 2018.

[205] P. Wang, W.H. Lin, K.M. Chao, C.C. Lo, A face-recognition approach using deep reinforcement learning approach for user authentication, in: 2017 IEEE 14th International Conference on e-Business Engineering (ICEBE), IEEE, 2017, pp. 183–188.

[206] Z. Wang, X. Tang, W. Luo, S. Gao, Face aging with identity-preserved conditional generative adversarial networks, in: Proceedings of the IEEE Conference on Computer Vision and Pattern Recognition, 2018, pp. 7939–7947.

[207] Y. Wen, B. Raj, R. Singh, Face reconstruction from voice using generative adversarial networks, in: Advances in Neural Information Processing Systems, 2019, pp. 5266–5275.

[208] Y. Wen, K. Zhang, Z. Li, Y. Qiao, A discriminative feature learning approach for deep face recognition, in: European Conference on Computer Vision, Springer, Cham, 2016, pp. 499–515.

[209] M. Wöllmer, A. Metallinou, F. Eyben, B. Schuller, S. Narayanan, Context-sensitive multimodal emotion recognition from speech and facial expression using bidirectional lstm modeling, in: Proc. INTERSPEECH 2010, Makuhari, Japan, 2010, pp. 2362–2365.

[210] W. Wu, C. Qian, S. Yang, Q. Wang, Y. Cai, Q. Zhou, Look at boundary: a boundary-aware face alignment algorithm, in: Proceedings of the IEEE Conference on Computer Vision and Pattern Recognition, 2018, pp. 2129–2138.

[211] X. Wu, R. He, Z. Sun, T. Tan, A light CNN for deep face representation with noisy labels, IEEE Transactions on Information Forensics and Security 13 (11) (2018) 2884–2896.

[212] Z. Wu, W. Deng, One-shot deep neural network for pose and illumination normalization face recognition, in: 2016 IEEE International Conference on Multimedia and Expo (ICME), IEEE, 2016, pp. 1–6.

[213] X. Xiong, F. De la Torre, Supervised descent method and its applications to face alignment, in: Proceedings of the IEEE Conference on Computer Vision and Pattern Recognition, 2013, pp. 532–539.

[214] Z. Xu, S. Li, W. Deng, Learning temporal features using LSTM-CNN architecture for face anti-spoofing, in: 2015 3rd IAPR Asian Conference on Pattern Recognition (ACPR), IEEE, 2015, pp. 141–145.

[215] B. Yang, J. Yan, Z. Lei, S.Z. Li, Aggregate channel features for multi-view face detection, in: IEEE International Joint Conference on Biometrics, IEEE, 2014, pp. 1–8.

[216] P. Yang, Q. Liu, D.N. Metaxas, Rankboost with l1 regularization for facial expression recognition and intensity estimation, in: 2009 IEEE 12th International Conference on Computer Vision, IEEE, 2009, pp. 1018–1025.

[217] S. Yang, P. Luo, C.C. Loy, X. Tang, Faceness-net: face detection through deep facial part responses, IEEE Transactions on Pattern Analysis and Machine Intelligence 40 (8) (2017) 1845–1859.

[218] S. Yang, Y. Xiong, C.C. Loy, X. Tang, Face detection through scale-friendly deep convolutional networks, arXiv preprint, arXiv:1706.02863, 2017.
[219] T.Y. Yang, Y.T. Chen, Y.Y. Lin, Y.Y. Chuang, FSA-Net: learning fine-grained structure aggregation for head pose estimation from a single image, in: Proceedings of the IEEE Conference on Computer Vision and Pattern Recognition, 2019, pp. 1087–1096.
[220] W. Yang, S. Li, W. Ouyang, H. Li, X. Wang, Learning feature pyramids for human pose estimation, in: Proceedings of the IEEE International Conference on Computer Vision, 2017, pp. 1281–1290.
[221] X. Yin, X. Yu, K. Sohn, X. Liu, M. Chandraker, Towards large-pose face frontalization in the wild, in: Proceedings of the IEEE International Conference on Computer Vision, 2017, pp. 3990–3999.
[222] Z. Yu, C. Zhang, Image based static facial expression recognition with multiple deep network learning, in: Proceedings of the 2015 ACM on International Conference on Multimodal Interaction, 2015, pp. 435–442.
[223] Z. Yu, G. Liu, Q. Liu, J. Deng, Spatio-temporal convolutional features with nested LSTM for facial expression recognition, Neurocomputing 317 (2018) 50–57.
[224] A. Zadeh, Y. Chong Lim, T. Baltrusaitis, L.P. Morency, Convolutional experts constrained local model for 3d facial landmark detection, in: Proceedings of the IEEE International Conference on Computer Vision Workshops, 2017, pp. 2519–2528.
[225] M.D. Zeiler, R. Fergus, Visualizing and understanding convolutional networks, in: European Conference on Computer Vision, Springer, Cham, 2014, pp. 818–833.
[226] B. Zhang, G. Liu, G. Xie, Facial expression recognition using LBP and LPQ based on Gabor wavelet transform, in: 2016 2nd IEEE International Conference on Computer and Communications (ICCC), IEEE, 2016, pp. 365–369.
[227] F. Zhang, T. Zhang, Q. Mao, C. Xu, Joint pose and expression modeling for facial expression recognition, in: Proceedings of the IEEE Conference on Computer Vision and Pattern Recognition, 2018, pp. 3359–3368.
[228] H. Zhang, T. Xu, H. Li, S. Zhang, X. Wang, X. Huang, D.N. Metaxas, Stackgan: text to photo-realistic image synthesis with stacked generative adversarial networks, in: Proceedings of the IEEE International Conference on Computer Vision, 2017, pp. 5907–5915.
[229] J. Zhang, S. Shan, M. Kan, X. Chen, Coarse-to-fine auto-encoder networks (cfan) for real-time face alignment, in: European Conference on Computer Vision, Springer, Cham, 2014, pp. 1–16.
[230] K. Zhang, Y. Huang, Y. Du, L. Wang, Facial expression recognition based on deep evolutional spatial-temporal networks, IEEE Transactions on Image Processing 26 (9) (2017) 4193–4203.
[231] N. Zhang, M. Paluri, M.A. Ranzato, T. Darrell, L. Bourdev, Panda: pose aligned networks for deep attribute modeling, in: Proceedings of the IEEE Conference on Computer Vision and Pattern Recognition, 2014, pp. 1637–1644.
[232] P. Michel, R. El Kaliouby, Real time facial expression recognition in video using support vector machines, in: 2003 Proceedings of the 5th International Conference on Multimodal Interfaces, 2003, pp. 258–264.
[233] S. Zhang, X. Zhu, Z. Lei, H. Shi, X. Wang, S.Z. Li, S3fd: single shot scale-invariant face detector, in: Proceedings of the IEEE International Conference on Computer Vision, 2017, pp. 192–201.
[234] X. Zhang, Z. Fang, Y. Wen, Z. Li, Y. Qiao, Range loss for deep face recognition with long-tailed training data, in: Proceedings of the IEEE International Conference on Computer Vision, 2017, pp. 5409–5418.
[235] Y. Zhang, L. Sun, Exploring correlations in multiple facial attributes through graph attention network, arXiv preprint, arXiv:1810.09162, 2018.

[236] F. Zhao, J. Feng, J. Zhao, W. Yang, S. Yan, Robust lstm-autoencoders for face de-occlusion in the wild, IEEE Transactions on Image Processing 27 (2) (2017) 778–790.
[237] X. Zhao, X. Liang, L. Liu, T. Li, Y. Han, N. Vasconcelos, S. Yan, Peak-piloted deep network for facial expression recognition, in: European Conference on Computer Vision, Springer, Cham, 2016, pp. 425–442.
[238] W. Zhen, Y. Zilu, Facial expression recognition based on adaptive local binary pattern and sparse representation, in: 2012 IEEE International Conference on Computer Science and Automation Engineering (CSAE), Vol. 2, IEEE, 2012, pp. 440–444.
[239] W. Zheng, H. Tang, Z. Lin, T.S. Huang, A novel approach to expression recognition from non-frontal face images, in: 2009 IEEE 12th International Conference on Computer Vision, IEEE, 2009, pp. 1901–1908.
[240] X. Zheng, Y. Guo, H. Huang, Y. Li, R. He, A survey of deep facial attribute analysis, arXiv preprint, arXiv:1812.10265, 2018.
[241] X. Zheng, H. Huang, Y. Guo, B. Wang, R. He, BLAN: bi-directional ladder attentive network for facial attribute prediction, Pattern Recognition 100 (2020) 107155.
[242] Y. Zheng, D.K. Pal, M. Savvides, Ring loss: convex feature normalization for face recognition, in: Proceedings of the IEEE Conference on Computer Vision and Pattern Recognition, 2018, pp. 5089–5097.
[243] L. Zhong, Q. Liu, P. Yang, B. Liu, J. Huang, D.N. Metaxas, Learning active facial patches for expression analysis, in: 2012 IEEE Conference on Computer Vision and Pattern Recognition, IEEE, 2012, pp. 2562–2569.
[244] Y. Zhong, J. Sullivan, H. Li, Leveraging mid-level deep representations for predicting face attributes in the wild, in: 2016 IEEE International Conference on Image Processing (ICIP), IEEE, 2016, pp. 3239–3243.
[245] Q. Zhu, M.C. Yeh, K.T. Cheng, S. Avidan, Fast human detection using a cascade of histograms of oriented gradients, in: 2006 IEEE Computer Society Conference on Computer Vision and Pattern Recognition (CVPR'06), vol. 2, IEEE, 2006, pp. 1491–1498.
[246] X. Zhu, Z. Lei, X. Liu, H. Shi, S.Z. Li, Face alignment across large poses: a 3d solution, in: Proceedings of the IEEE Conference on Computer Vision and Pattern Recognition, 2016, pp. 146–155.
[247] N. Zhuang, Y. Yan, S. Chen, H. Wang, Multi-task learning of cascaded CNN for facial attribute classification, in: 2018 24th International Conference on Pattern Recognition (ICPR), 2018, pp. 2069–2074.
[248] M. Suwa, N. Sugie, K. Fujimora, A preliminary note on pattern recognition of human emotional expression, in: Proceedings of the Fourth International Joint Conference on Pattern Recognition, Kyoto, 1978.

CHAPTER 6

Finite multi-dimensional generalized Gamma Mixture Model Learning for feature selection

Basim Alghabashi[a], Mohamed Al Mashrgy[b], Muhammad Azam[c], and Nizar Bouguila[a]
[a]Concordia Institute for Information Systems Engineering (CIISE), Concordia University, Montreal, QC, Canada
[b]Electrical and Computer Engineering (ECE), Al-Mergib University, Alkhums, Libya
[c]Electrical and Computer Engineering (ECE), Concordia University, Montreal, QC, Canada

Contents

6.1 Introduction

Classification and clustering analysis are prevalent in different disciplines that require data analysis, such as computer vision, image processing, pattern recognition, medicine and machine learning [1–5]. Many clustering techniques can fall under two categories: First, heuristic algorithms where no probabilistic models are explicitly assumed (e.g. K-Means), and model-based is the second approach, which makes inferences via assumptions of the data distribution [6]. Moreover, supervised and unsupervised learning are different model-based techniques that machine learning follows [7,8].

Learning Control
https://doi.org/10.1016/B978-0-12-822314-7.00011-0

Finite-mixture models are a powerful and flexible tool used for unsupervised learning to cluster univariate and multivariate data [9–11]. Furthermore, finite-mixture models are used to draw inferences from unlabeled observations in order to perform the clustering analysis. The necessity of analyzing multi-dimensional data is increasing day after day, which plays an important role in exploiting finite statistical model-based approaches in order to infer useful knowledge [12–16]. Model selection and selecting the relevant features at the same time, is a daunting task in particular when positive multi-dimensional data are considered [17].

Good results have been obtained when utilizing Gaussian mixture models with various applications such as human and car detection, infrared face recognition and texture clustering [18–20], but on the other hand, non-Gaussian ones have recently shown promising results that outperform Gaussian distributions, which is due to the fact that the shape of Gaussian distributions is rigid. Some examples include, but are not limited to finite multi-dimensional generalized Gamma and Gamma mixtures [21–27]. It is noteworthy to point out the significant merit that generalized Gamma mixture model (GGMM) has which is the flexibility of its shape compared to Gamma and Gaussian models.

One of the crucial issues that has been widely studied is the determination of the optimal number of clusters that best describes the data. As a solution, the minimum message length (MML) criterion is adopted and derived to tackle this and obtain the required model's complexity [28]. Related works have demonstrated appealing outcomes when exploiting MML criterion with different applications such as classification of Web pages, texture database summarization for efficient retrieval, SAR image analysis, shadow detection in images and summarization of texture images for efficient retrieval, and handwriting recognition [29–35].

Another problem to take into consideration is the presence of too many features that maybe irrelevant which could affect the performance and accuracy of the learning model [36]. Tackling this issue, will help preventing over-fitting, speeding up the performance and increasing the accuracy of the model [23,37]. In a previous work, a framework based on a finite multi-dimensional generalized Gamma mixture model was developed to cluster positive vectors and determine the number of clusters which best describes the data [21].

In this paper, we extend the previous work [21] by considering feature selection which in turn performs the task of drawing inferences from unlabeled observations to cluster multi-dimensional positive vectors that are naturally generated by many applications [17]. To the best of our knowledge, feature selection when deploying generalized Gamma mixture models for multi-dimensional data has never been considered in the past. What makes the proposed model stand out, when comparing it to the rest of the related research works, is its capability of dealing with multi-dimensional positive vectors. The one-dimensional generalized Gamma was considered in [38], [39], and [40] with its applications in Ultrasonic tissue characterization, statistical modeling of high-resolution and blind signal separation (BSS), respectively. In addition, it was first introduced by Stacty [41] in 1962 which was achieved by introducing a positive exponential parameter in Gamma distribution.

The rest of this paper is organized as follows. In Section 6.2 we present the proposed generalized Gamma mixture model. In Section 6.3 the parameters estimation algorithm is explained. The formulation of the MML criterion in the case of the generalized Gamma mixture is presented in Section 6.4, and then Section 6.5 evaluates the proposed model performance. Finally, the last section is devoted to the conclusion and future works.

6.2 The proposed model

Suppose we have a data set $\mathcal{X} = \{\vec{X}_1, \vec{X}_2, ..., \vec{X}_n\}$ where each $\vec{X}_i = (X_{i1}, X_{i2}, ..., X_{iD})$ is a D-dimensional positive vector. Suppose that these vectors follow a mixture of multi-dimensional generalized Gamma distributions, where M represents the number of components, and $i = 1, ..., N$. Therefore, the following is the mixture model:

$$p(\vec{X}_i|\Theta_M) = \sum_{j=1}^{M} p_j p(\vec{X}_i|\theta_j) \tag{6.1}$$

where $\Theta_M = \{\theta_1, \theta_2, ..., \theta_M, p_1, ..., p_M\}$, and θ_j is a set of the distribution parameters of class j, where $\theta_j = \{\vec{\alpha}_j, \vec{\beta}_j, \vec{\lambda}_j\}$, with $\vec{\alpha}_j = (\alpha_{j1}, \alpha_{j2}, ..., \alpha_{jD})$, $\vec{\beta}_j = (\beta_{j1}, \beta_{j2}, ..., \beta_{jD})$, and $\vec{\lambda}_j = (\lambda_{j1}, \lambda_{j2}, ..., \lambda_{jD})$. In addition, the scale, shape and extra shape parameters are α_{jd}, β_{jd} and λ_{jd}, respectively. The parameter p_j is the mixing proportion such that $0 \leqslant p_j \leqslant 1$, and $\sum_{j=1}^{M} p_j = 1$. We have

$$p(\vec{X}_i|\theta_j) = \prod_{d=1}^{D} p(X_{id}|\theta_{jd}) = \prod_{d=1}^{D} \frac{\frac{\lambda_{jd}}{\alpha_{jd}^{\beta_{jd}}} X_{id}^{\beta_{jd}-1} \exp - \left(\frac{X_{id}}{\alpha_{jd}}\right)^{\lambda_{jd}}}{\Gamma(\frac{\beta_{jd}}{\lambda_{jd}})} \tag{6.2}$$

where $\theta_{jd} = (\alpha_{jd}, \beta_{jd}, \lambda_{jd})$ and $\Gamma(.)$ denotes the Gamma function, and $d = 1, \ldots, D$. The likelihood can be expressed by

$$p(\mathcal{X}|\Theta) = \prod_{i=1}^{N} \sum_{j=1}^{M} p_j p(\vec{X}_i|\theta_j). \tag{6.3}$$

Assume that $Z_i = (Z_{i1}, Z_{i2}, ..., Z_{iM})$ are the membership vectors, which determines the cluster j that each observation belongs to, such that $Z_{ij} \in \{0, 1\}$, and $Z_{ij} = 1$ if $\vec{X}_i$ belongs to cluster j and 0, otherwise. The complete-data likelihood function is given as follows:

$$p(\mathcal{X}, Z|\Theta) = \prod_{i=1}^{N} \prod_{j=1}^{M} p_j p(\vec{X}_i|\theta_j)^{Z_{ij}}. \tag{6.4}$$

6.3 Parameter estimation

As mentioned above, feature selection or determining which attributes to use when conducting the clustering task is a crucial step, because some features are just noise that affects the accuracy of the model. The complete log-likelihood function when integrating feature selection can be expressed by

$$\begin{aligned}\log p(\mathcal{X}, Z|\Theta) = \sum_{i=1}^{N} \sum_{j=1}^{M} Z_{ij} \log p_j \prod_{d=1}^{D} \hat{\rho}_{jd} p(\vec{X}_{id}|\theta_j) \\ + (1 - \hat{\rho}_{jd}) p(X_{id}|\hat{\tau}_{jd}).\end{aligned} \tag{6.5}$$

The main goal is to maximize the log-likelihood function by taking the gradient with the respect to each parameter p_j, α_{jd}, β_{jd}, and λ_{jd}. $\hat{\rho}_{jd}$ represents the weight of the dth feature on cluster j, and $p(X_{id}|\hat{\tau}_{jd})$, $\tau_{jd} = (\alpha_{\tau|jd}, \beta_{\tau|jd}, \lambda_{\tau|jd})$ is the generalized Gamma distribution in which the irrelevant feature is discarded. As a result we obtain the function given by

$$L(\mathcal{X}, Z|\Theta) = \sum_{i=1}^{N} \sum_{j=1}^{M} Z_{ij} \log p_j \hat{\rho}_{jd} p(\vec{X}_i|\theta_j) + (1 - \hat{\rho}_{jd}) p(X_{id}|\hat{\tau}_{jd}). \tag{6.6}$$

By differentiating the previous function w.r.t. p_j, α_{jd}, β_{jd} and λ_{id}, we get

$$p_j = \frac{\sum_{i=1}^{N} p(j|\vec{X}_i)}{N} \tag{6.7}$$

where p_j is the prior probability. We have

$$\hat{Z}_{ij} = p(j|\vec{X}_i) = \frac{p_j \prod_{d=1}^{D} \left(\rho_{jd} p(X_{id}|\theta_{jd}) + (1-\rho_{jd}) p(X_{id}|\tau_{jd})\right)}{\sum_{j=1}^{M} \left(p_j \prod_{d=1}^{D} \left(\rho_{jd} p(X_{id}|\theta_{jd}) + (1-\rho_{jd}) p(X_{id}|\tau_{jd})\right)\right)}, \quad (6.8)$$

which is the posterior probability which its main role is assigning a vector $\vec{X}_i$ to a particular cluster $j, j = 1, \ldots, M$.

$$\rho_{jd} = \frac{\sum_{i=1}^{N} p(j|\vec{X}_i) f(\rho_{jd}, \theta_{jd}, \tau_{jd})}{\sum_{i=1}^{N} p(j|\vec{X}_i)} \quad j = 1, \ldots, M \quad d = 1, \ldots, D, \quad (6.9)$$

$$f(\rho_{jd}, \theta_{jd}, \tau_{jd}) = \frac{\rho_{jd} p(X_{id}|\theta_{jd})}{\rho_{jd} p(X_{id}|\theta_{jd}) + (1-\rho_{jd}) \rho_{jd} p(X_{id}|\tau_{jd})}, \quad (6.10)$$

where $f(\rho_{jd}, \theta_{jd}, \rho_{jd})$ is the posterior probability of a given feature d relevancy.

$$f(1-\rho_{jd}, \theta_{jd}, \tau_{jd}) = \frac{(1-\rho_{jd}) p(X_{id}|\tau_{jd})}{\rho_{jd} p(X_{id}|\theta_{jd}) + (1-\rho_{jd}) \rho_{jd} p(X_{id}|\tau_{jd})}, \quad (6.11)$$

$f(1-\rho_{jd}, \theta_{jd}, \tau_{jd})$ is the posterior probability for the irrelevant feature d. We have

$$\alpha_{jd} = \left(\frac{\sum_{i=1}^{N} p(j|\vec{X}_i) f(\rho_{jd}, \theta_{jd}, \tau_{jd}) \left(\lambda_{jd} X_{id}^{\lambda_{jd}}\right)}{\sum_{i=1}^{N} p(j|\vec{X}_i) \beta_{jd} f(\rho_{jd}, \theta_{jd}, \tau_{jd})} \right)^{\frac{1}{\lambda_{jd}}}, \quad (6.12)$$

$$\alpha_{\tau|jd} = \left(\frac{\sum_{i=1}^{N} p(j|\vec{X}_i) f(1-\rho_{jd}, \theta_{jd}, \tau_{jd}) \left(\lambda_{jd} X_{id}^{\lambda_{jd}}\right)}{\sum_{i=1}^{N} p(j|\vec{X}_i) \beta_{jd} f(1-\rho_{jd}, \theta_{jd}, \tau_{jd})} \right)^{\frac{1}{\lambda_{jd}}}, \quad (6.13)$$

$$\beta_{jd} = \frac{\lambda_{jd} \Psi^{-1}}{\sum_{i=1}^{N} p(j|\vec{X}_i) f(\rho_{jd}, \theta_{jd}, \tau_{jd})} \times \left(\lambda_{jd} \left(\sum_{i=1}^{N} p(j|\vec{X}_i) f(\rho_{jd}, \theta_{jd}, \tau_{jd}) \left[-\log(\alpha_{jd}) + \log(X_{id}) \right] \right) \right), \quad (6.14)$$

$$\beta_{\tau|jd} = \frac{\lambda_{jd}\Psi^{-1}}{\sum_{i=1}^{N} p(j|\vec{X}_i) f(1-\rho_{jd}, \theta_{jd}, \tau_{jd})} \times \left(\lambda_{jd}\left(\sum_{i=1}^{N} p(j|\vec{X}_i) f(1-\rho_{jd}, \theta_{jd}, \tau_{jd})\left[-\log(\alpha_{jd}) + \log(X_{id})\right]\right)\right), \tag{6.15}$$

where $\Psi^{-1}(.)$ is the inverse digamma function. By taking the gradient of the log-likelihood w.r.t. λ_{jd}, we obtain

$$\frac{\partial \log p(\mathcal{X}, Z|\Theta)}{\partial \lambda_{jd}} = \sum_{i=1}^{N} p(j|\vec{X}_i) f(\rho_{jd}, \theta_{jd}, \tau_{jd}) \times \left[\frac{1}{\lambda_{jd}} - \left((\frac{X_{id}}{\alpha_{jd}})^{\lambda_{jd}} \log(\frac{X_{id}}{\alpha_{jd}})\right) + \frac{\Psi(\frac{\beta_{jd}}{\lambda_{jd}})\beta_{jd}}{\lambda_{jd}^2}\right]. \tag{6.16}$$

Looking at Eq. (6.16), we can see that we do not have a linear equation. As a solution, we have used the Newton–Raphson method in [26] to solve this problem and estimate λ_{jd}, which can be expressed as follows:

$$\lambda_{jd}^{new} = \lambda_{jd}^{old} - \gamma \frac{\partial \log p(\mathcal{X}|\Theta)}{\partial \lambda_{jd}} \left(\frac{\partial^2 \log p(\mathcal{X}|\Theta)}{\partial^2 \lambda_{jd}}\right)^{-1} \tag{6.17}$$

where γ is the constant step size added to Newton's method. After computing $\frac{\partial^2 \log p(\mathcal{X}|\Theta)}{\partial^2 \lambda_{jd}}$ we get

$$\frac{\partial p(\mathcal{X}|\Theta)}{\partial \lambda_{jd}^2} = \sum_{i=1}^{N} p(j|\vec{X}_i) f(\rho_{jd}, \theta_{jd}, \tau_{jd}) \left[-\frac{1}{\lambda_{jd}^2} - (\frac{X_{id}}{\alpha_{jd}})^{\lambda_{jd}} \log(\frac{X_{id}}{\alpha_{jd}}) \log(\frac{X_{id}}{\alpha_{jd}}) - \left(\frac{\psi'(\frac{\beta_{jd}}{\lambda_{jd}})\beta^2}{\lambda_{jd}^4}\right) - \left(\frac{\beta_{jd} 2\psi(\frac{\beta_{jd}}{\lambda_{jd}})}{\lambda_{jd}^3}\right)\right]$$

where $\psi'(.)$ is the trigamma function.

6.4 Model selection using the minimum message length criterion

Choosing the optimal number of components in order to accurately represent the data is a challenging task and critical step when exploiting mixture models [42–49]. As a solution, MML is utilized to develop a criterion that is capable of determining the number of clusters of generalized Gamma distributions:

$$MessLen(M) \simeq -log(h(\Theta)) - log(p(\mathcal{X}|\Theta)) + \frac{1}{2}log(|F(\Theta)|) + \frac{N_p}{2}(1 - log(k_{Np})) \quad (6.18)$$

where $h(\Theta)$ is the prior probability, $p(\mathcal{X}|\Theta)$ is the likelihood, $F(\Theta)$ is the expected Fisher information matrix, $|F(\Theta)|$ is its determinant, and N_p is number of free parameters to be estimated which is equal to $M(D+2)-1$. k_{Np} is the optimal quantization lattice constant for $\mathbb{R}^{N_p}$ and we have $k_1 = 1/12 \simeq 0.083$ for $N_p = 1$ [29]. Our main goal is to compute the minimum value with regards to Θ of the message length (*MessLen*) so that we can find the optimal number of components. In the subsequent subsections, we first compute the Fisher information for a mixture of generalized Gamma distributions and then we chose the proper prior distributions.

6.4.1 Fisher information for a generalized Gamma mixture model

We computed the Fisher information by taking the determinant of the mixture model Hessian matrix. The determinant of the complete-data Fisher information matrix that has a block-diagonal structure can be expressed as follows:

$$|F(\vec{\Theta})| \approx |F(\vec{P})| \prod_{d=1}^{D} |F(\rho_{jd})||F(\vec{\alpha}_{jd})||F(\vec{\beta}_{jd})||F(\vec{\lambda}_{jd})|F(\vec{\alpha}_{\tau|jd})| |F(\vec{\beta}_{\tau|jd})||F(\vec{\lambda}_{\tau|jd})|, \quad (6.19)$$

where $|F(\vec{P})|$ is the determinant of the Fisher information with regard to the mixing parameters vector that must satisfy the constraint $\sum_{j=1}^{M} p_j = 1$ [29]. $|F(\rho_{jd})|$, $|F(\vec{\alpha}_{jd})|$, $|F(\vec{\beta}_{jd})|$, $|F(\vec{\lambda}_{jd})|$, $F(\vec{\alpha}_{\tau|jd})|$, $|F(\vec{\beta}_{\tau|jd})|$, and $|F(\vec{\lambda}_{\tau|jd})|$ are the Fisher information with regards to the vectors $\vec{\alpha}_{jd}$, $\vec{\beta}_{jd}$, $\vec{\lambda}_{jd}$, respectively, of a single generalized Gamma distribution. Consequently, it is possible to consider the generalized Bernoulli process with a series of trials, each of which has M possible outcomes labeled first cluster, second cluster, ..., Mth cluster. Thus, the number of trials of jth cluster is a multinomial distribution of parameters $p_1, p_2, \ldots, p_M$. Hence, the determinant of the Fisher information matrix [50]:

$$|F(\vec{P})| = \frac{N^{M-1}}{\prod_{j=1}^{M} p_j} \tag{6.20}$$

where N is the number of elements in cluster j, and M is the number of clusters. The Hessian matrices when we consider the vectors $\vec{\alpha}_{jd}$, $\vec{\beta}_{jd}$, $\vec{\lambda}_{jd}$, $\vec{\alpha}_{\tau|jd}$, $\vec{\beta}_{\tau|jd}$, and $\vec{\lambda}_{\tau|jd}$ are given by

$$F(\vec{\rho}_j)_{d_1 d_2} = \frac{\partial^2}{\partial \rho_{jd_1} \rho_{jd_2}} \{log[p(\mathcal{X}|\Theta)]\}, \tag{6.21}$$

$$F(\vec{\alpha}_j)_{d_1 d_2} = \frac{\partial^2}{\partial \alpha_{jd_1} \alpha_{jd_2}} \{log[p(\mathcal{X}|\Theta)]\}, \tag{6.22}$$

$$F(\vec{\beta}_j)_{d_1 d_2} = \frac{\partial^2}{\partial \beta_{jd_1} \beta_{jd_2}} \{log[p(\mathcal{X}|\Theta)]\}, \tag{6.23}$$

$$F(\vec{\lambda}_j)_{d_1 d_2} = \frac{\partial^2}{\partial \lambda_{jd_1} \lambda_{jd_2}} \{log[p(\mathcal{X}|\Theta)]\}, \tag{6.24}$$

$$F(\vec{\alpha}_{\tau|j})_{d_1 d_2} = \frac{\partial^2}{\partial \alpha_{\tau|jd_1} \alpha_{\tau|jd_2}} \{log[p(\mathcal{X}|\Theta)]\}, \tag{6.25}$$

$$F(\vec{\beta}_{\tau|j})_{d_1 d_2} = \frac{\partial^2}{\partial \beta_{\tau|jd_1} \beta_{\tau|jd_2}} \{log[p(\mathcal{X}|\Theta)]\}, \tag{6.26}$$

$$F(\vec{\lambda}_{\tau|j})_{d_1 d_2} = \frac{\partial^2}{\partial \lambda_{\tau|jd_1} \lambda_{\tau|jd_2}} \{log[p(\mathcal{X}|\Theta)]\}, \tag{6.27}$$

where $(d_1, d_2) \in (1, 2, \ldots .D)$. By computing the derivatives in Eqs. (6.22), (6.23), (6.24), we obtain

$$|F(\rho_{jd})| = \sum_{i=1}^{N} p(j|X_i)^2 \left(\frac{f(1-\rho_{jd}, \theta_d, \tau_d)}{1-\rho_{jd}} - \frac{f(\rho_{jd}, \theta_d, \tau_d)}{\rho_{jd}} \right)^2 \tag{6.28}$$

$$|F(\vec{\alpha}_j)| = -\sum_{i=1}^{N} p(j|\vec{X}_i) f(\rho_{jd}, \theta_{jd}, \tau_{jd}) \left[\frac{\beta_{jd}}{\alpha_{jd}^2} + (-\lambda_{jd} - 1)\alpha^{-\lambda_{jd}-2} \lambda_{jd} X_{id}^{\lambda_{jd}} \right], \tag{6.29}$$

$$|F(\vec{\beta}_j)| = -\sum_{i=1}^{N} p(j|\vec{X}_i) f(\rho_{jd}, \theta_{jd}, \tau_{jd}) \left[-\frac{\psi'(\frac{\beta_{jd}}{\lambda_{jd}})}{\lambda_{jd}^2} \right], \tag{6.30}$$

$$|F(\vec{\lambda}_j)| = -\sum_{i=1}^{N} p(j|\vec{X}_i) f(\rho_{jd}, \theta_{jd}, \tau_{jd}) \left[-\frac{1}{\lambda_{jd}^2} - (\frac{X_{id}}{\alpha_{jd}})^{\lambda_{jd}} \right.$$

$$\left. \log(\frac{X_{id}}{\alpha_{jd}}) \log(\frac{X_{id}}{\alpha_{jd}}) - \left(\frac{\psi'(\frac{\beta_{jd}}{\lambda_{jd}})\beta^2}{\lambda_{jd}^4} \right) - \left(\frac{\beta_{jd} 2\psi(\frac{\beta_{jd}}{\lambda_{jd}})}{\lambda_{jd}^3} \right) \right]. \tag{6.31}$$

6.4.2 Prior distribution $h(\Theta)$

The prior distributions for the parameters α_{jd} and β_{jd} for GGMM are the same ones used for generalized Gamma distribution [23],

$$h(\Theta) = h(\vec{P}) \prod_{j=1}^{M} \prod_{d=1}^{D} h(\rho_{jd}) h(\alpha_{jd}) h(\beta_{jd}) h(\lambda_{jd})$$

$$h(\alpha_{\tau|jd}) h(\beta_{\tau|jd}) h(\lambda_{\tau|jd}); \tag{6.32}$$

it is known that the vector $\vec{P}$ is defined in the simplex $\{(p(1), \ldots, p(M)) : \sum_{j=1}^{M} p(j) = 1\}$, then it is common to assume a symmetric Dirichlet distribution as a prior distribution [29],

$$h(\vec{p}) = \frac{\Gamma(\sum_{j=1}^{M} \eta_j)}{\prod_{j=1}^{M} \Gamma(\eta_j)} \prod_{j=1}^{M} p_j^{\eta_j - 1} \tag{6.33}$$

where $\vec{\eta} = (\eta_1, \ldots, \eta_M)$ is the parameter vector of the Dirichlet distribution. The choice of $\eta = 1$ gives us the prior:

$$h(\vec{p}) = \frac{1}{(M-1)!}. \tag{6.34}$$

A symmetric Beta distribution is assumed to be a prior for the parameter ρ_{jd}, and this is because it is known that ρ_{jd} is defined on the compact support $[0, 1]$, and setting the parameters equal to 1 gives the uniform prior $h(\rho_{jd}) = U_{[0,1]}$. The proper priors for the model parameters θ are selected depending on related research works and results obtained from our experiments. For the scale parameters α_{jd} and $\alpha_{\tau|jd}$ we chose the following priors [23]:

$$h(\alpha_{jd}) = \frac{1}{\alpha_{jd}}, \tag{6.35}$$

$$h(\alpha_{\tau|jd}) = \frac{1}{\alpha_{\tau|jd}}, \tag{6.36}$$

and for the shape parameters β_{jd}, and $\beta_{\tau|jd}$ we chose exponential priors [23]:

$$h(\beta_{jd}) = 10^{-2}exp(-10^{-2}\beta_{jd}), \quad (6.37)$$

$$h(\beta_{\tau|jd}) = 10^{-2}exp(-10^{-2}\beta_{\tau|jd}), \quad (6.38)$$

for the extra shape parameters λ_{jd} and $\lambda_{\tau|jd}$ we followed the prior distribution used in [51], which adopted a uniform distribution $U[0, h]$:

$$h(\lambda_{jd}) = \frac{1}{h^{M.D}}, \quad (6.39)$$

$$h(\lambda_{\tau|jd}) = \frac{1}{h^{M.D}}. \quad (6.40)$$

6.4.3 Algorithm

As mentioned before, the learning algorithm is based on expectation maximization method which has a complexity of $O(NMD)$. For the initialization of the EM algorithm, we used K-Means and Method of Moments (MoM) techniques.

6.5 Experimental results

In this section the effectiveness of the proposed method is tested using real-life multi-dimensional data sets, that include texture, shape and scene images, respectively.

6.5.1 Texture images

Texture analysis and modeling is of great importance in image analysis and its applications. *Vistex* is a well-known data set that contains texture images and obtained from the Massachusetts Institute of Technology (MIT) Media Lab.[1] We divided each mother image into sub-images following the same method in [52]. The database consists of four homogeneous texture groups, "bark", "fabric", "sand", and "water". We have 64 images (64 × 64 pixels each) for each group obtained by dividing each 256 × 256 mother image into sub-images. Features for each sub-image are extracted using Histogram of Oriented Gradients (HOG), which gives us an 81-dimensional positive vector. Fig. 6.1 shows examples of images from the *Vistex* data set.

[1] http://vismod.media.mit.edu/vismod/imagery/VisionTexture/distribution.html.

Algorithm 1 Multi-dimensional generalized Gamma Mixture Parameters Estimation.

procedure
  INPUT: D-Dimensional data set $\mathcal{X} = \{X_1, X_2, \ldots, X_n\}$, t_{min}, M_{max}
  OUTPUT: Θ^*, M^*
  for $M = 1 : M_{max}$ **do**
    Initialization algorithm:
    applying K-Means on N D-dimensional vectors
    computing p_j
    obtaining α_j and β_j by applying the method of moments
    obtaining λ_j by using random positive values.
    EM algorithm:
    while relative change in log-likelihood $\geq t_{min}$ **do**
      E-Step:
      compute the posterior probabilities $P(j|\vec{X}_i)$ using Eq. (6.8)
      M-Step:
      update p_j using Eq. (6.7)
      update ρ_{jd} using Eq. (6.9)
      update α_{jd} using Eq. (6.12)
      update $\alpha_{t|jd}$ using Eq. (6.13)
      update β_{jd} using Eq. (6.14)
      update $\beta_{t|jd}$ using Eq. (6.15)
      start Newton's Raphson Algorithm
      for all $1 \leq j \leq M$ **do**
        update λ_j using Eq. (6.17)
      end for
      end Newton's Raphson Algorithm
    end while
    calculate the associated criterion MML (M) using Eq. (6.17)
  end for
  Select the optimal model M^* such that $M^* = argmin_M$ MML(M)
end procedure

We used generalized Gamma, Gamma and Gaussian mixture models in order to perform texture clustering. To show the importance of feature selection, we have conducted the first experiment without considering the relevancy of features. Table 6.1 shows the confusion matrix which illustrates the results when using the proposed model without feature selection.

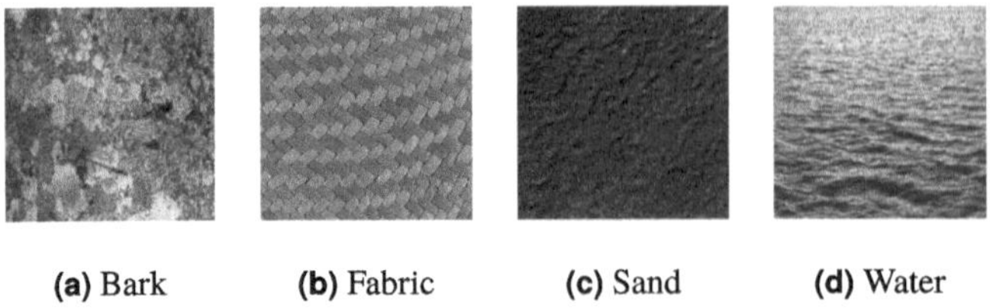

Figure 6.1 Samples from *Vistex* data set.

Table 6.1 Confusion matrix for the texture clustering using a generalized Gamma mixture model without taking into consideration the relevancy of features.

	Bark	Fabric	Sand	Water
Bark	64	0	0	0
Fabric	0	64	0	0
Sand	0	2	62	0
Water	0	0	0	64

Having two misclassified sand images gives us 99.21% clustering accuracy. On the other hand, the clustering results when using Gamma mixture are shown in Table 6.2, and as we can see there are three misclassified images with 98.82% accuracy. In addition, the result when using a Gaussian mixture model is shown in Table 6.3. There are 10 misclassified images, which gives us 96.09% accuracy.

Table 6.2 Confusion matrix for the texture clustering using a Gamma mixture model without taking into consideration the relevancy of features.

	Bark	Fabric	Sand	Water
Bark	64	0	0	0
Fabric	2	61	1	0
Sand	0	0	64	0
Water	0	0	0	64

The second experiment on texture images is conducted when taking into consideration the relevancy of features. Tables 6.4 and 6.5 show the results when using generalized Gamma and Gamma distributions, respectively. One misclassified texture image using GGMM represents 99.6% clustering accuracy, while when using Gamma, the number of misclassi-

Table 6.3 Confusion matrix for texture the clustering using a Gaussian mixture model without taking into consideration the relevancy of features.

	Bark	Fabric	Sand	Water
Bark	64	0	0	0
Fabric	1	61	2	0
Sand	0	0	64	0
Water	0	7	0	57

Table 6.4 Confusion matrix for the texture clustering using a generalized Gamma mixture model when taking into consideration the relevancy of features.

	Bark	Fabric	Sand	Water
Bark	64	0	0	0
Fabric	0	64	0	0
Sand	1	0	63	0
Water	0	0	0	64

Table 6.5 Confusion matrix for the texture clustering using a Gamma mixture model when taking into consideration the relevancy of features.

	Bark	Fabric	Sand	Water
Bark	64	0	0	0
Fabric	0	63	0	1
Sand	1	0	63	0
Water	0	0	0	64

fied textures is two, which represents an accuracy of 99.21%. Moreover, Table 6.6 shows the results when using a Gaussian distribution. The number of misclassified images is eight, which represents 96.87% accuracy. The accuracy of the three models is increased when enabling the feature selection. For example, the accuracy of GGMM is improved by 0.39%. Also, GGMM always acquires the highest accuracy compared to Gamma and Gaussian mixtures. Fig. 6.2 shows the saliency of features for the generalized Gamma mixture. Each feature is shown with its standard deviation if existed. Also, different weight is assigned to each feature (between 0 and 1) which represents its significance/relevancy. Moreover, Fig. 6.3 shows us

Table 6.6 Confusion matrix for the texture clustering using a Gaussian mixture model when taking into consideration the relevancy of features.

	Bark	Fabric	Sand	Water
Bark	64	0	0	0
Fabric	0	64	0	0
Sand	0	3	59	2
Water	3	0	0	61

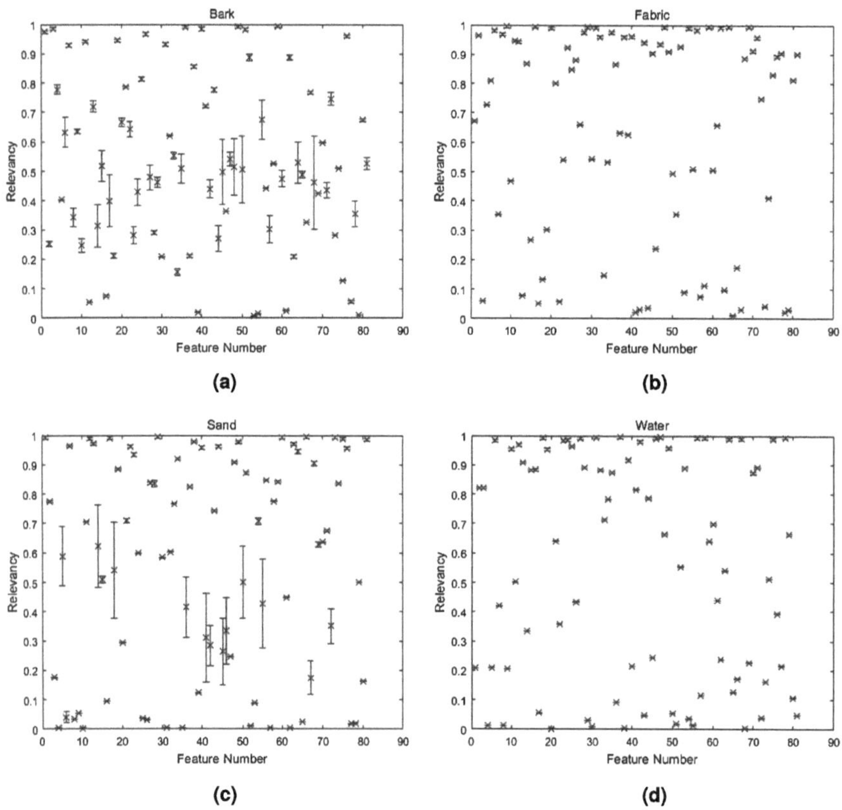

Figure 6.2 Features saliency for the *Vistex* data set using GGMM.

that the optimal number of clusters when using MML criterion is four, which was based on the minimum message length.

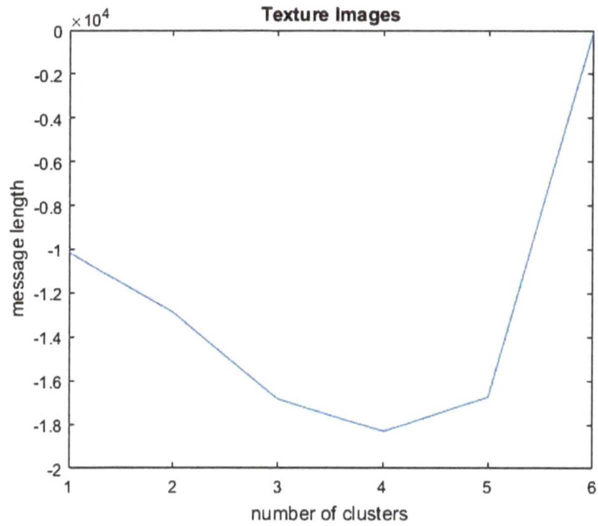

Figure 6.3 Optimal number of clusters for the texture images data set using MML criterion.

6.5.2 Shape images

In this section, experiments are conducted to test the proposed model performance, using a well-known real data set to solve the issue of clustering multi-dimensional positive vectors, feature selection and finding the optimal number of components that best describes the data. We have used seven classes from MPEG-7 CE Shape-1 Part-B data set,[2] where each one has 20 samples (140 images in total). For each image, a 36-dimensional positive vector of features is extracted using Zernike Moments Magnitudes. Zernike Moments Magnitudes is an effective method used to extract features from shape images [23]. Examples of its applications include invariant image recognition [53], and region-based shape modeling [54]. Moreover, we followed the method in [55] in order to extract the positive vector of features, which imposes the constraints such that $n = 0, 1, 2, ..., \infty$, $|m| \leq n$, and $n - |m|$ is even, where n is the order of the Zernike moments and m is the Zernike moment repetition number. Fig. 6.4 shows some examples from the data set, and Tables 6.7, and 6.8 illustrate the confusion matrices for generalized Gamma mixture without and with taking into consideration the relevancy of features, respectively. The accuracy of the model has been

[2] http://www.daibi.temple.edu/~shape/MPEG7/dataset.html.

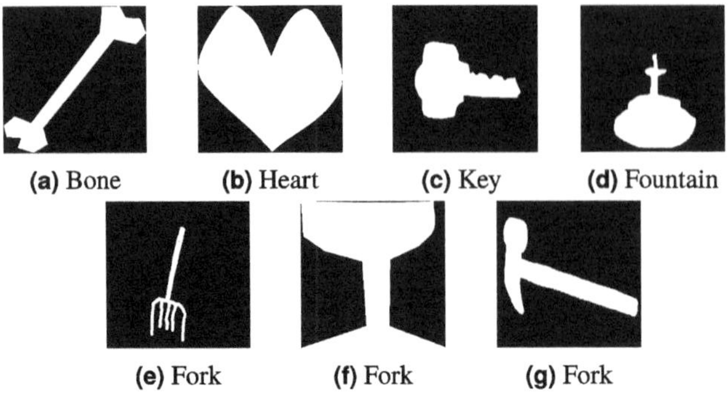

Figure 6.4 Samples from MPEG-7 CE Shape-1 Part-B.

Table 6.7 Confusion matrix for the shape images using a generalized Gamma mixture model without taking into consideration the relevancy of features.

	Bones	Hearts	Keys	Fountains	Forks	Glass	Hummer
Bones	19	0	0	0	0	1	0
Hearts	0	19	1	0	0	0	0
Keys	0	1	19	0	0	0	0
Fountains	1	0	0	14	0	3	2
Forks	0	0	0	0	20	0	0
Glass	1	0	0	8	0	11	0
Hummer	1	0	0	0	0	0	19

Table 6.8 Confusion matrix for the shape images using a generalized Gamma mixture model when taking into consideration the relevancy of features.

	Bones	Hearts	Keys	Fountains	Forks	Glass	Hummer
Bones	20	0	0	0	0	0	0
Hearts	0	19	1	0	0	0	0
Keys	0	1	19	0	0	0	0
Fountains	0	0	1	14	0	3	2
Forks	0	0	0	0	20	0	0
Glass	1	0	0	8	0	11	0
Hummer	1	0	0	0	0	0	19

significantly increased from 86.42% (19 misclassified images) to 87.14% (18 misclassified images), and this is due to the fact that irrelevant features can affect the model accuracy. Fig. 6.5 shows the features saliency of the shape images data set obtained by using a generalized Gamma mixture model. Each image is represented by 36 features, and each one has a different

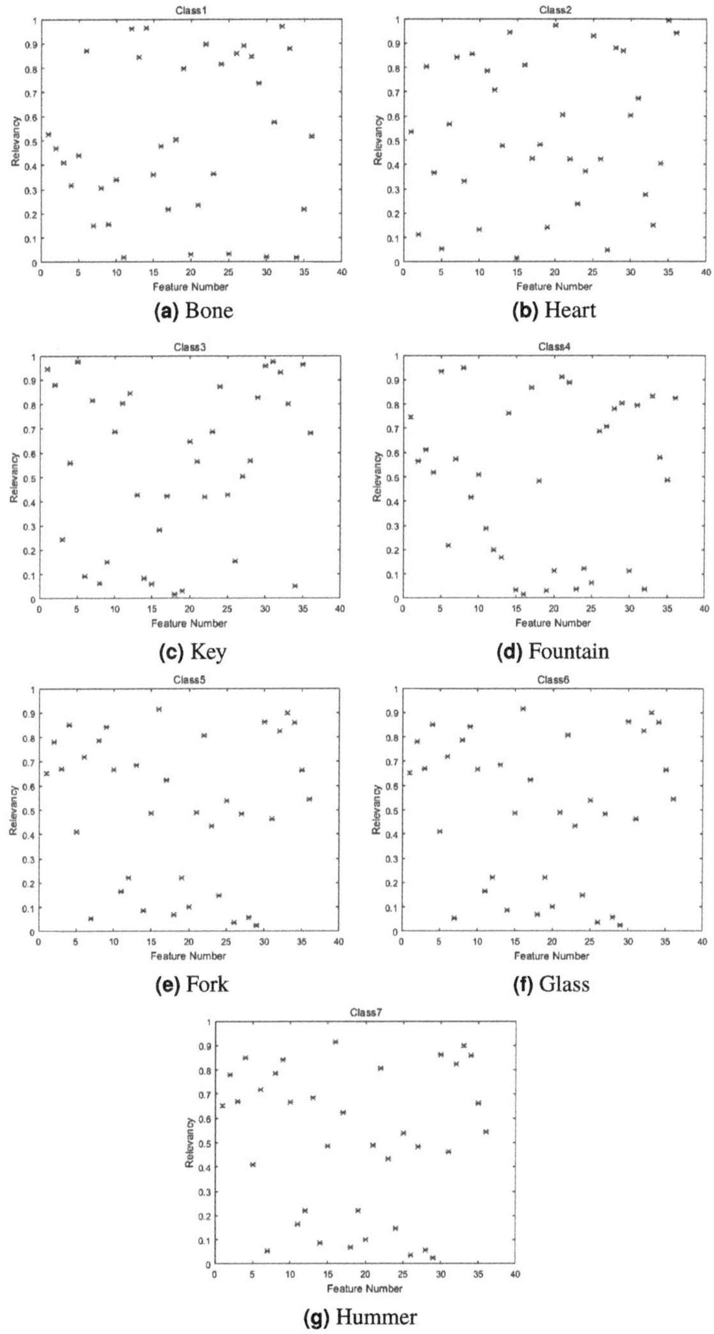

Figure 6.5 Features saliency for the MPEG-7 CE Shape-1 Part-B data set using GGMM.

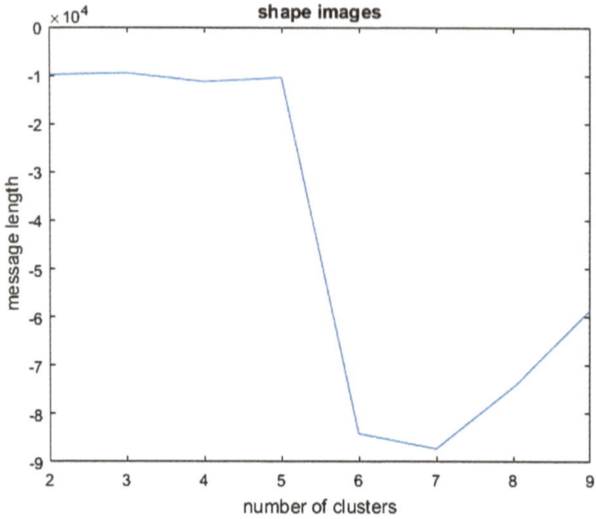

Figure 6.6 Optimal number of clusters for the shape images data set using MML criteria.

weight, which helps discriminating between significant and insignificant features. For example, in Fig. 6.5 (b), feature number 15 is completely irrelevant, while feature number 35 is highly significant. The results of the obtained number of clusters for the shape images data set when using MML criterion are shown in Fig. 6.6. The obtained results lead us to the minimum message length, which shows that the proposed model is performing well and capable of finding the optimal number of clusters ($M = 7$).

The second experiment is conducted using Gamma mixtures without and with taking into consideration the relevancy of features, respectively. Tables 6.9, and 6.10 show the accuracy of Gamma distribution which has been improved from 82.86% to 85% when considering the feature selection. The last experiment is conducted using Gaussian mixtures without and with taking into consideration the relevancy of features, respectively. Tables 6.11 and 6.12 show the accuracy of Gaussian distribution which has been improved from 65.51% to 72.41% when considering the feature selection (these results are obtained from [23]). To summarize, using feature selection has a significant role to play in increasing the clustering accuracy, and by comparing the results of the three models, namely generalized Gamma, Gamma, and Gaussian distributions, we can see that GGMM always has the highest accuracy.

Table 6.9 Confusion matrix for the shape images using a Gamma mixture model without taking into consideration the relevancy of features.

	Bones	Hearts	Keys	Fountains	Forks	Glass	Hummer
Bones	16	0	0	0	0	4	0
Hearts	0	18	0	2	0	0	0
Keys	0	0	17	0	0	0	3
Fountains	0	1	3	16	0	0	0
Forks	1	4	0	3	14	0	1
Glass	0	2	0	0	1	17	0
Hummer	2	0	0	0	0	0	18

Table 6.10 Confusion matrix for the shape images using a Gamma mixture model when taking into consideration the relevancy of features.

	Bones	Hearts	Keys	Fountains	Forks	Glass	Hummer
Bones	20	0	0	0	0	0	0
Hearts	0	19	1	0	0	0	0
Keys	0	1	19	0	0	0	0
Fountains	1	0	0	12	0	3	4
Forks	0	0	0	0	19	0	1
Glass	1	0	0	8	0	11	0
Hummer	1	0	0	0	0	0	19

Table 6.11 Confusion matrix for the shape images using a Gaussian mixture model without taking into consideration the relevancy of features.

	Bones	Hearts	Keys	Fountains	Forks	Glass	Hummer
Bones	19	1	0	0	0	0	0
Hearts	4	16	0	0	0	0	0
Keys	0	6	10	0	0	2	2
Fountains	0	0	0	14	0	3	3
Forks	5	0	0	0	15	0	0
Glass	1	6	0	0	0	13	0
Hummer	12	0	0	0	0	0	8

6.5.3 Scene images

Scene clustering is another challenging application used to test the performance of the proposed algorithm. We used the same data set in [6], which was obtained from [56]. Five classes are used (living room, forest, coast, tall building, and mountain), where each one has 56 samples (280 images in total). Features for each image are extracted using HOG, which gives us an 81-dimensional positive vector. Fig. 6.7 shows examples from the

Table 6.12 Confusion matrix for the shape images using a Gaussian mixture model when taking into consideration the relevancy of features.

	Bones	Hearts	Keys	Fountains	Forks	Glass	Hummer
Bones	14	0	0	1	0	0	5
Hearts	0	19	0	0	1	0	0
Keys	0	0	16	0	0	0	4
Fountains	0	8	0	12	0	0	0
Forks	0	0	0	3	17	0	0
Glass	0	0	0	0	2	18	0
Hummer	2	0	0	0	5	4	9

Table 6.13 Confusion matrix for the visual scene images using a generalized Gamma mixture model without taking into consideration the relevancy of features.

	Living room	Forest	Coast	Tall building	Mountain
Living room	45	0	9	0	45
Forest	0	39	0	0	0
Coast	1	0	55	0	1
Tall building	2	0	19	20	2
Mountain	50	0	0	4	50

Table 6.14 Confusion matrix for the visual scene images using a Gamma mixture model without taking into consideration the relevancy of features.

	Living room	Forest	Coast	Tall building	Mountain
Living room	48	0	2	2	4
Forest	0	42	0	14	0
Coast	1	0	55	0	0
Tall building	0	0	25	31	0
Mountain	24	1	0	6	25

scene images data set. First experiment is conducted without the consideration of feature selection. Table 6.13 shows the confusion matrix which illustrates the results when using GGMM without feature selection. Having 71 misclassified images represents 74.64% clustering accuracy. On the other hand, Tables 6.14 and 6.15 show the clustering results when using gamma (79 misclassified images with 71.78% clustering accuracy) and Gaussian (85 misclassified images with 69.64% clustering accuracy) mixtures, respectively. The second experiment on scene images is conducted when taking into consideration the relevancy of features. Fig. 6.8 shows the saliency of features obtained by using the generalized gamma mixture model, which represents the weight/relevancy of each feature with its stan-

Figure 6.7 Samples from the five classes scene images data set.

Table 6.15 Confusion matrix for the visual scene images using a Gaussian mixture model without taking into consideration the relevancy of features.

	Living room	Forest	Coast	Tall building	Mountain
Living room	43	2	3	1	7
Forest	3	47	1	5	0
Coast	0	0	48	7	1
Tall building	2	5	22	23	4
Mountain	10	3	5	4	34

dard deviation. Fig. 6.9 shows the results of the obtained number of clusters for the scene images data set when using MML. As a result, the minimum message length leads us to the correct number of clusters ($M = 5$). Tables 6.16 and 6.17 show the results when using generalized Gamma and Gamma distributions, respectively. 65 misclassified images represent 76.79% clustering accuracy when using GGMM, while when using Gamma mixture, the number of misclassified images is 71, which represents 74.64% clustering accuracy. Moreover, Table 6.18 shows the results when using a

Table 6.16 Confusion matrix for the visual scene images using a generalized Gamma mixture model when taking into consideration the relevancy of features.

	Living room	Forest	Coast	Tall building	Mountain
Living room	51	0	0	0	5
Forest	0	49	0	7	0
Coast	0	0	52	2	2
Tall building	0	1	17	32	6
Mountain	20	0	2	3	31

Table 6.17 Confusion matrix for the visual scene images using a Gamma mixture model when taking into consideration the relevancy of features.

	Living room	Forest	Coast	Tall building	Mountain
Living room	51	0	0	0	5
Forest	0	46	0	10	0
Coast	0	0	52	2	2
Tall building	0	1	20	29	6
Mountain	19	0	3	3	31

Table 6.18 Confusion matrix for the visual scene images using a Gaussian mixture model when taking into consideration the relevancy of features.

	Living room	Forest	Coast	Tall building	Mountain
Living room	46	3	2	4	1
Forest	3	40	2	4	7
Coast	6	13	32	4	1
Tall building	1	3	7	43	2
Mountain	0	8	3	8	37

Gaussian distribution. The result shows that the number of misclassified images is 82, which represents 70.71% clustering accuracy. Indeed, the use of feature selection allows to increase the clustering accuracy for all the models. For example, the clustering accuracy of GGMM is improved by 2.15%, and as we can see, GGMM always obtains the highest accuracy.

6.6 Conclusion

Finite-mixture models are useful and powerful tool used for clustering data. In this paper, a finite generalized Gamma mixture model is presented in order to cluster unlabeled positive multi-dimensional vectors, determining the number of clusters, and feature selection simultaneously. The model parameters have been estimated by conducting ML approach via EM. The

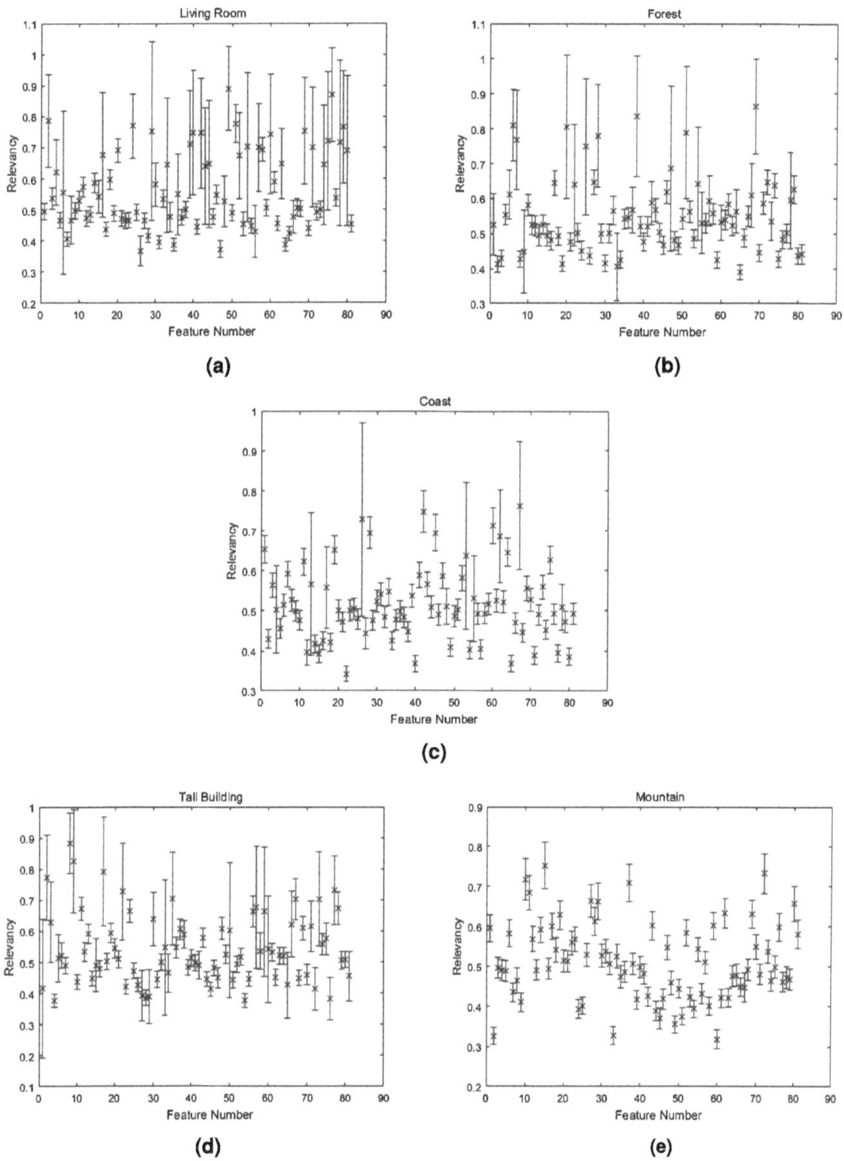

Figure 6.8 Features saliency for the five classes scene images data set using GGMM.

main goal is to increase the model accuracy by taking into consideration the feature relevancy. The flexibility of the shape is one of the outstanding properties that gives this model the ability to achieve outstanding results.

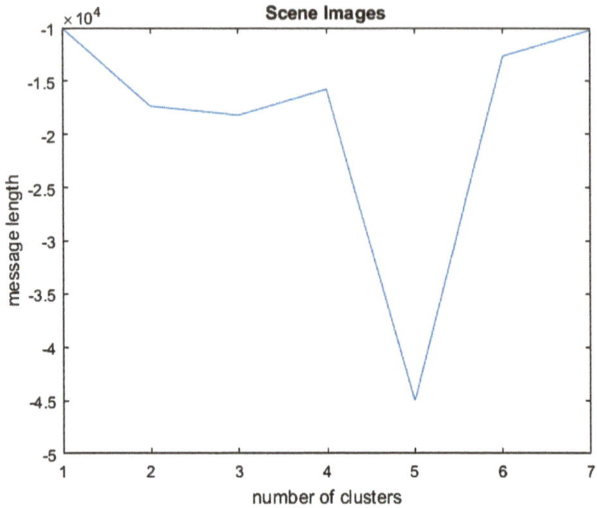

Figure 6.9 Optimal number of clusters for the five classes scene images data set using MML criteria.

Due to this fact, the proposed methodology gives a result that outperforms other ones such as Gaussian and Gamma mixture models. The model has been validated by using three real-life applications, namely textures, shapes and scenes clustering. Thus, results show that the proposed model clustering accuracy shows the high performance capability of GGMM. One of the avenues for future work is to develop a Bayesian learning approach using infinite statistical mixture models to tackle the issue of clustering high-dimensional dynamic data.

References

[1] Abir Hussain, Machine learning approaches for extracting genetic medical data information, in: Proceedings of the Second International Conference on Internet of Things and Cloud Computing, ICC 2017, Cambridge, United Kingdom, March 22–23, 2017, 2017, p. 1.

[2] Mohammed Khalaf, Abir Jaafar Hussain, Dhiya Al-Jumeily, Russell Keenan, Paul Fergus, Ibrahim Olatunji Idowu, Robust approach for medical data classification and deploying self-care management system for sickle cell disease, in: 15th IEEE International Conference on Computer and Information Technology; Ubiquitous Computing and Communications, Dependable, Autonomic and Secure Computing, Pervasive Intelligence and Computing, 2015, pp. 575–580.

[3] Ala S. Al Kafri, Sud Sudirman, Abir Jaafar Hussain, Paul Fergus, Dhiya Al-Jumeily, Mohammed Al-Jumaily, Haya Al-Askar, A framework on a computer assisted and systematic methodology for detection of chronic lower back pain using artificial intelligence and computer graphics technologies, in: Intelligent Computing Theories and

Application – 12th International Conference, ICIC 2016, Lanzhou, China, August 2–5, 2016, Proceedings, Part I, 2016, pp. 843–854.

[4] A.K. Jain, M.N. Murty, P.J. Flynn, Data clustering: a review, ACM Computing Surveys 31 (3) (September 1999) 264–323.

[5] Leonid Portnoy, Eleazar Eskin, Sal Stolfo, Intrusion detection with unlabeled data using clustering, in: Proceedings of ACM CSS Workshop on Data Mining Applied to Security, DMSA-2001, 2001, pp. 5–8.

[6] Mohamed Al Mashrgy, Taoufik Bdiri, Nizar Bouguila, Robust simultaneous positive data clustering and unsupervised feature selection using generalized inverted Dirichlet mixture models, Knowledge-Based Systems 59 (2014) 182–195.

[7] Xiaojin Zhu, Semi-supervised learning literature survey, Technical Report 1530, Computer Sciences, University of Wisconsin–Madison, 2005.

[8] Nizar Bouguila, Djemel Ziou, Using unsupervised learning of a finite Dirichlet mixture model to improve pattern recognition applications, Pattern Recognition Letters 26 (12) (2005) 1916–1925.

[9] Geoffrey McLachlan, Finite Mixture Models, Wiley, New York, 2000.

[10] O. Amayri, N. Bouguila, Infinite Langevin mixture modeling and feature selection, in: 2016 IEEE International Conference on Data Science and Advanced Analytics (DSAA), Oct 2016, pp. 149–155.

[11] M.A.T. Figueiredo, A.K. Jain, Unsupervised learning of finite mixture models, IEEE Transactions on Pattern Analysis and Machine Intelligence 24 (3) (March 2002) 381–396.

[12] Zeynep Tufekci, Big Questions for Social Media Big Data: Representativeness, Validity and Other Methodological Pitfalls, The AAAI Press, 2014, pp. 505–514.

[13] N. Bouguila, Spatial color image databases summarization, in: 2007 IEEE International Conference on Acoustics, Speech and Signal Processing – ICASSP '07, Vol. 1, 2007, pp. I-953–I-956.

[14] N. Bouguila, Count data modeling and classification using finite mixtures of distributions, IEEE Transactions on Neural Networks 22 (2) (2011) 186–198.

[15] N. Bouguila, W. ElGuebaly, A generative model for spatial color image databases categorization, in: 2008 IEEE International Conference on Acoustics, Speech and Signal Processing, 2008, pp. 821–824.

[16] Nizar Bouguila, Walid ElGuebaly, On discrete data clustering, in: Takashi Washio, Einoshin Suzuki, Kai Ming Ting, Akihiro Inokuchi (Eds.), Advances in Knowledge Discovery and Data Mining, 12th Pacific-Asia Conference, PAKDD 2008, Osaka, Japan, May 20–23, 2008 Proceedings, in: Lecture Notes in Computer Science, vol. 5012, Springer, 2008, pp. 503–510.

[17] G.V. Trunk, A problem of dimensionality: a simple example, IEEE Transactions on Pattern Analysis and Machine Intelligence PAMI-1 (3) (July 1979) 306–307.

[18] Mohamed Al Mashrgy, Nizar Bouguila, A fully Bayesian framework for positive data clustering, in: Artificial Intelligence Applications in Information and Communication Technologies, 2015, pp. 147–164.

[19] Tarek Elguebaly, Nizar Bouguila, Infinite generalized Gaussian mixture modeling and applications, in: Mohamed Kamel, Aurélio Campilho (Eds.), Image Analysis and Recognition, Springer, Berlin, Heidelberg, 2011, pp. 201–210.

[20] M. Azam, N. Bouguila, Unsupervised keyword spotting using bounded generalized Gaussian mixture model with ICA, in: 2015 IEEE Global Conference on Signal and Information Processing (GlobalSIP), Dec 2015, pp. 1150–1154.

[21] Basim Alghabashi, Nizar Bouguila, Finite multi-dimensional generalized gamma mixture model learning based on MML, in: 17th IEEE International Conference on Machine Learning and Applications, ICMLA 2018, Orlando, FL, USA, December 17–20, 2018, 2018, pp. 1131–1138.

[22] Basim Alghabashi, Nizar Bouguila, A finite multi-dimensional generalized gamma mixture model, in: The 2018 IEEE International Conference on Smart Data (SmartData-2018), Halifax, Canada, Jul 2018.
[23] Nizar Bouguila, Khaled Almakadmeh, Sabri Boutemedjet, A finite mixture model for simultaneous high-dimensional clustering, localized feature selection and outlier rejection, Expert Systems with Applications 39 (7) (2012) 6641–6656.
[24] F.R. Al-Osaim, N. Bouguila, A finite gamma mixture model-based discriminative learning frameworks, in: 2015 IEEE 14th International Conference on Machine Learning and Applications (ICMLA), Dec 2015, pp. 819–824.
[25] D. Ziou, N. Bouguila, Unsupervised learning of a finite gamma mixture using MML: application to SAR image analysis 2 (Aug 2004) 68.
[26] Taoufik Bdiri, Nizar Bouguila, Positive vectors clustering using inverted Dirichlet finite mixture models, Expert Systems with Applications 39 (2) (2012) 1869–1882.
[27] Ali Sefidpour, Nizar Bouguila, Spatial color image segmentation based on finite non-Gaussian mixture models, Expert Systems with Applications 39 (10) (2012) 8993–9001.
[28] C.S. Wallace, D.M. Boulton, An information measure for classification, Computer Journal 11 (2) (1968) 185–194.
[29] N. Bouguila, D. Ziou, High-dimensional unsupervised selection and estimation of a finite generalized Dirichlet mixture model based on minimum message length, IEEE Transactions on Pattern Analysis and Machine Intelligence 29 (10) (Oct 2007) 1716–1731.
[30] D. Ziou, N. Bouguila, Unsupervised learning of a finite gamma mixture using MML: application to SAR image analysis, in: Proceedings of the 17th International Conference on Pattern Recognition, ICPR 2004, Vol. 2, Aug 2004, pp. 68–71.
[31] N. Bouguila, D. Ziou, Unsupervised selection of a finite Dirichlet mixture model: an MML-based approach, IEEE Transactions on Knowledge and Data Engineering 18 (8) (Aug 2006) 993–1009.
[32] Nizar Bouguila, Djemel Ziou, On fitting finite Dirichlet mixture using ECM and MML, in: Sameer Singh, Maneesha Singh, Chid Apte, Petra Perner (Eds.), Pattern Recognition and Data Mining, Springer, Berlin, Heidelberg, 2005, pp. 172–182.
[33] Nizar Bouguila, Djemel Ziou, MML-based approach for finite Dirichlet mixture estimation and selection, in: Machine Learning and Data Mining in Pattern Recognition, 2005, pp. 42–51.
[34] N. Bouguila, D. Ziou, MML-based approach for high-dimensional unsupervised learning using the generalized Dirichlet mixture, in: 2005 IEEE Computer Society Conference on Computer Vision and Pattern Recognition (CVPR'05) – Workshops, June 2005, p. 53.
[35] Nuha Zamzami, Nizar Bouguila, MML-based approach for determining the number of topics in EDCM mixture models, in: Ebrahim Bagheri, Jackie C.K. Cheung (Eds.), Advances in Artificial Intelligence, Springer International Publishing, Cham, 2018, pp. 211–217.
[36] Keinosuke Fukunaga, Introduction to Statistical Pattern Recognition, 2nd ed., Academic Press Professional, Inc., San Diego, CA, USA, 1990.
[37] Q. Hu, W. Pedrycz, D. Yu, J. Lang, Selecting discrete and continuous features based on neighborhood decision error minimization, IEEE Transactions on Systems, Man and Cybernetics Part B Cybernetics 40 (1) (Feb 2010) 137–150.
[38] Gonzalo Vegas-Sanchez-Ferrero, Santiago Aja-Fernandez, Cesar Palencia, Marcos Martin-Fernandez, A generalized gamma mixture model for ultrasonic tissue characterization, Computational and Mathematical Methods in Medicine 2012 (2012) 1–25.
[39] H.C. Li, V.A. Krylov, P.Z. Fan, J. Zerubia, W.J. Emery, Unsupervised learning of generalized gamma mixture model with application in statistical modeling of high-resolution SAR images, IEEE Transactions on Geoscience and Remote Sensing 54 (4) (April 2016) 2153–2170.

[40] M. El-Sayed Waheed, Osama Abdo Mohamed, M.E. Abd El-Aziz, Mixture of generalized gamma density-based score function for FastICA, Mathematical Problems in Engineering 2011 (2011) 1–14.

[41] E.W. Stacy, A generalization of the gamma distribution, The Annals of Mathematical Statistics 33 (3) (Sep 1962) 1187–1192.

[42] Nizar Bouguila, Djemel Ziou, Dirichlet-based probability model applied to human skin detection [image skin detection], in: 2004 IEEE International Conference on Acoustics, Speech, and Signal Processing, ICASSP 2004, Montreal, Quebec, Canada, May 17–21, 2004, 2004, pp. 521–524.

[43] N. Bouguila, D. Ziou, A Dirichlet process mixture of Dirichlet distributions for classification and prediction, in: 2008 IEEE Workshop on Machine Learning for Signal Processing, 2008, pp. 297–302.

[44] N. Bouguila, D. Ziou, Improving content based image retrieval systems using finite multinomial Dirichlet mixture, in: Proceedings of the 2004 14th IEEE Signal Processing Society Workshop Machine Learning for Signal Processing, 2004, pp. 23–32.

[45] Nizar Bouguila, Djemel Ziou, A powerful finite mixture model based on the generalized Dirichlet distribution: unsupervised learning and applications, in: 17th International Conference on Pattern Recognition, ICPR 2004, Cambridge, UK, August 23–26, 2004, 2004, pp. 280–283.

[46] Nizar Bouguila, Bayesian hybrid generative discriminative learning based on finite Liouville mixture models, Pattern Recognition 44 (6) (2011) 1183–1200.

[47] Tarek Elguebaly, Nizar Bouguila, Finite asymmetric generalized Gaussian mixture models learning for infrared object detection, Computer Vision and Image Understanding 117 (12) (2013) 1659–1671.

[48] Wentao Fan, Nizar Bouguila, Variational learning of a Dirichlet process of generalized Dirichlet distributions for simultaneous clustering and feature selection, Pattern Recognition 46 (10) (2013) 2754–2769.

[49] Nizar Bouguila, Djemel Ziou, A countably infinite mixture model for clustering and feature selection, Knowledge and Information Systems 33 (2) (2012) 351–370.

[50] Rohan A. Baxter, Jonathan J. Oliver, Finding overlapping components with MML, Statistics and Computing 10 (1) (Jan 2000) 5–16.

[51] Djemel Ziou Mohand Said Allili, Nizar Bouguila, Finite general Gaussian mixture modeling and application to image and video foreground segmentation, Journal of Electronic Imaging 17 (2008) 17.

[52] Nizar Bouguila, Djemel Ziou, Unsupervised learning of a finite discrete mixture: applications to texture modeling and image databases summarization, Journal of Visual Communication and Image Representation 18 (4) (2007) 295–309.

[53] A. Khotanzad, Y.H. Hong, Invariant image recognition by Zernike moments, IEEE Transactions on Pattern Analysis and Machine Intelligence 12 (5) (May 1990) 489–497.

[54] Whoi-Yul Kim, Yong-Sung Kim, A region-based shape descriptor using Zernike moments, Signal Processing Image Communication 16 (1) (2000) 95–102.

[55] Y.S. Kim, W.Y. Kim, Content-based trademark retrieval system using a visually salient feature, Image and Vision Computing 16 (12) (1998) 931–939.

[56] Aude Oliva, Antonio Torralba, Modeling the shape of the scene: a holistic representation of the spatial envelope, International Journal of Computer Vision 42 (3) (May 2001) 145–175.

CHAPTER 7

Variational learning of finite shifted scaled Dirichlet mixture models

Zeinab Arjmandiasl, Narges Manouchehri, Nizar Bouguila, and Jamal Bentahar
Concordia Institute for Information Systems Engineering (CIISE), Concordia University, Montreal, QC, Canada

Contents

7.1 Introduction

Data mining and dealing with large and complex data has become an important part of decision-making procedures in various research domains. Clustering as a powerful statistical approach has been effectively and extensively applied in finding hidden patterns within data [1], [2]. Among all clustering and unsupervised learning methods, finite mixture models specifically have shown remarkable success in various applications [3], [4], [5]. The principal idea of this clustering technique is to fit a mixture of

Learning Control
https://doi.org/10.1016/B978-0-12-822314-7.00012-2

components derived from a predetermined distribution into the data via parameter estimation of the components [6], [7], [8].

There are two conventional methods to learn finite mixture models, namely deterministic and Bayesian approaches. Each of these methods has certain downsides [6]. For instance, deterministic approaches such as maximum likelihood estimation (MLE) do not produce a decent approximation given a small dataset, plus overfitting, being sensitive to initialization and converging to local maxima instead of the global one [9], [10], [11]. Bayesian approaches [12], [13] with the incorporation of prior knowledge can overcome the aforementioned problems through simulation techniques, but they have their own drawbacks [14], such as being computationally complex and time consuming specially for high dimensional data [15], [16].

A proposed alternative to avoid such disadvantages is the variational Bayesian approach. In this framework, parameters are modeled by assuming an approximation for their true posterior and minimizing the Kullback–Leibler (KL) divergence between approximated and true posterior [13], [17]. Variational learning has shown higher performance while being less computationally expensive in comparison to a traditional Bayesian approach [17], [18]. Another advantage of a variational framework is its capability of finding the correct number of components automatically along with a parameter estimation process. This is outstanding when we consider the MLE approach alone failing to find model complexity and its need for a model selection criterion such as a minimum message length [19]. Furthermore, the variational inference approach has proven to be superior to ML in parameter estimation [20], [21], [22], [23].

In mixture models, choosing a proper distribution to represent data is a pivotal step in order to achieve outstanding results. The Gaussian mixture model (GMM) has been a center of interest in most of previous data analysis researches due to its simplicity of estimation procedures [24]. However, GMMs are not always the best solution for any type of data specially for non-Gaussian ones. Recent researches have proven the prominent capability of other distributions such as Dirichlet distributions over the well-known GMM in many applications particularly when dealing with proportional data [25], [26], [19]. The aforementioned work, which has been done using maximum likelihood (ML) within the EM framework, has shown the flexibility and effectiveness of the Dirichlet distribution family, in particular, a generalized version of Dirichlet named the shifted-scaled Dirichlet distribution [27]. In this chapter, we propose variational learning of a mixture model based on shifted-scaled Dirichlet distribution which has two extra parameters, scale and location to make it more flexible and capable to spread out.

The rest of the chapter is structured as follows. In Section 7.2, the specifications of shifted-scaled Dirichlet finite mixture models are introduced. In Section 7.3, we discuss the steps of variational learning approach. The experimental results are presented in Section 7.4. Finally, Section 7.5 concludes our work.

7.2 Model specification

In this section, first we present shifted-scaled Dirichlet distribution. Afterward, the construction of mixture model based on this distribution will be explained.

7.2.1 Shifted-scaled Dirichlet distribution

A generalized version of Dirichlet distribution studied from a probabilistic point of view is the shifted-scaled Dirichlet distribution. This random composition is derived by applying two operations in the simplex, perturbation and powering. A vector-space structure is defined by these operations which play the same role as the sum and product by scalars in real space [28]. This added set of parameters has been shown to attain many functional probability models [29] which can be employed to model compositional multivariate data.

Let us assume an observation, generated from a shifted-scaled Dirichlet distribution (SSD), which is defined by $\vec{X} = (X_1, \ldots, X_D)$ as a random vector of proportional data where $\sum_{d=1}^{D} X_d = 1$, $0 \leq X_d \leq 1$. The parameters of this distribution are $\vec{\alpha} = (\alpha_1, \ldots, \alpha_D) \in \mathbb{R}_+^D$, $\vec{\beta} = (\beta_1, \ldots, \beta_D) \in \mathbb{S}^D$ and $\tau \in \mathbb{R}_+$. This distribution is expressed as follows:

$$p(\vec{X} \mid \theta) = \frac{\Gamma(\alpha+)}{\prod_{d=1}^{D} \Gamma(\alpha_d)} \frac{1}{\tau^{D-1}} \frac{\prod_{d=1}^{D} \beta_d^{-\frac{\alpha_d}{\tau}} X_d^{(\frac{\alpha_d}{\tau}-1)}}{\left(\sum_{d=1}^{D} (\frac{X_d}{\beta_d})^{\frac{1}{\tau}}\right)^{\alpha+}} \tag{7.1}$$

where $\Gamma(.)$ denotes the gamma function, $\vec{\alpha}$ is the shape parameter which represents the form of the distribution and $\alpha_+ = \sum_{d=1}^{D} \alpha_d$. $\vec{\beta}$ is the location parameter which refers to the data densities location and τ is a real scalar which tunes the variance of the density plot [27]. These parameters make our probability distribution remarkably flexible which empowers our model to fit various kinds of datasets.

7.2.2 Finite shifted-scaled Dirichlet mixture model

A convex combination of two or more probability distributions is called finite mixture model. Consider a set of N independent identically distributed observations described by $\mathcal{X} = (\vec{X}_1, \ldots, \vec{X}_N)$ in which each sample is a D-dimensional vector, $\vec{X}_i = (X_{i1}, \ldots, X_{iD})$ assumed to be generated from Eq. (7.2). We assume that this dataset could be explained by a finite mixture model including M components as follows [30]:

$$p\left(\vec{X}_i \mid \Theta\right) = \sum_{j=1}^{M} \pi_j p(\vec{X}_i \mid \vec{\theta}_j) \tag{7.2}$$

where π_j is the mixing coefficient of component j satisfying two constraints $0 < \pi_j < 1$, $\sum_{j=1}^{M} \pi_j = 1$.

$\Theta = \{\pi_1, \ldots, \pi_M, \vec{\theta}_1, \ldots, \vec{\theta}_M\}$ denotes the complete set of model parameters in which $\vec{\theta}_j = \{\vec{\alpha}_j, \vec{\beta}_j, \vec{\tau}_j\}$ represents the parameter vector for the jth component. Therefore the likelihood function of SSD mixture model is given by

$$p(\mathcal{X} \mid \vec{\pi}, \vec{\theta}) = \prod_{i=1}^{N} \left\{ \sum_{j=1}^{M} \pi_j p(\vec{X}_i \mid \vec{\theta}_j) \right\}. \tag{7.3}$$

For each $\vec{X}_i$, we introduce a M-dimensional random vector $\vec{Z}_i = (Z_{i1}, \ldots, Z_{iM})$ where $Z_{ij} \in \{0, 1\}$, $\sum_{j=1}^{M} Z_{ij} = 1$. This latent variable is not directly observed in the model, but from this we infer the cluster to which $\vec{X}_i$ is assigned such that $Z_{ij} = 1$ if it belongs to cluster j and 0 otherwise.

Therefore, the conditional probability distribution for the N hidden variables $\mathcal{Z} = (\vec{Z}_1, \ldots, \vec{Z}_N)$ given $\vec{\pi}$ is defined as

$$p(\mathcal{Z} \mid \vec{\pi}) = \prod_{j=1}^{N} \prod_{j=1}^{M} \pi_j^{Z_{ij}}. \tag{7.4}$$

Thus, the conditional probability of dataset $\mathcal{X}$ given the class labels $\mathcal{Z}$ is as follows where $\vec{\alpha} = (\vec{\alpha}_1, \ldots, \vec{\alpha}_M)$, $\vec{\beta} = (\vec{\beta}_1, \ldots, \vec{\beta}_M)$ and $\vec{\tau} = (\tau_1, \ldots, \tau_M)$:

$$p(\mathcal{X} \mid \mathcal{Z}, \vec{\alpha}, \vec{\beta}, \vec{\tau}) = \prod_{i=1}^{N} \prod_{j=1}^{M} \left[p(\vec{X}_i \mid \vec{\theta}_j)\right]^{Z_{ij}}. \tag{7.5}$$

7.3 Variational Bayesian learning

One of the crucial stages in fitting a dataset is learning the parameters of the mixture model that includes both parameter estimation and detection of number of components (M). In this section, we introduce a variational Bayesian approach for learning of shifted-scaled Dirichlet mixture models (varSSDMMs) which attain both of the above-mentioned problems simultaneously.

7.3.1 Parameter estimation

The joint distribution of all the random variables conditioned on $\vec{\pi}$ is given by

$$p(\mathcal{X}, \Theta \mid \vec{\pi}) = p\left(\mathcal{X} \mid \mathcal{Z}, \vec{\alpha}, \vec{\beta}, \vec{\tau}\right) p(\mathcal{Z} \mid \vec{\pi}) p(\vec{\alpha}) p(\vec{\beta}) p(\vec{\tau}) \tag{7.6}$$

where $\Theta = \{\mathcal{Z}, \vec{\alpha}, \vec{\beta}, \vec{\tau}\}$ and the following conjugate priors are chosen for $\vec{\alpha}, \vec{\beta}, \vec{\tau}$, respectively:

$$p(\alpha_{jd}) = \mathcal{G}(\alpha_{jd} \mid u_{jd}, v_{jd}) = \frac{v_{jd}^{u_{jd}}}{\Gamma(u_{jd})} \alpha_{jd}^{u_{jd}-1} e^{-v_{jd}\alpha_{jd}}, \tag{7.7}$$

$$p(\beta_{jd}) = \mathcal{D}(\beta_{jd} \mid \vec{h}_j) = \frac{\Gamma(\sum_{d=1}^{D} h_{jd})}{\prod_{d=1}^{D} \Gamma(h_{jd})} \prod_{d=1}^{D} \beta_{jd}^{h_{jd}-1}, \tag{7.8}$$

$$p(\tau_j) = \mathcal{G}(\tau_j \mid q_j, s_j) = \frac{q_j^{s_j}}{\Gamma(q_j)} \tau_j^{q_j-1} e^{-s_j\tau_j}; \tag{7.9}$$

$\{u_{jd}\}$, $\{v_{jd}\}$, $\{h_{jd}\}$, $\{q_{jd}\}$ and $\{s_{jd}\}$ are hyper-parameters which all satisfy the constraint of being greater than zero. $\mathcal{G}(.)$ and $\mathcal{D}(.)$ denote Gamma and Dirichlet distributions, respectively. Since the parameters are considered to be statistically independent, we can write

$$p(\vec{\alpha}) = \prod_{j=1}^{M} \prod_{d=1}^{D} p(\alpha_{jd}), \tag{7.10}$$

$$p(\vec{\beta}) = \prod_{j=1}^{M} \prod_{d=1}^{D} p(\beta_{jd}), \tag{7.11}$$

$$p(\vec{\tau}) = \prod_{j=1}^{M} p(\tau_j). \tag{7.12}$$

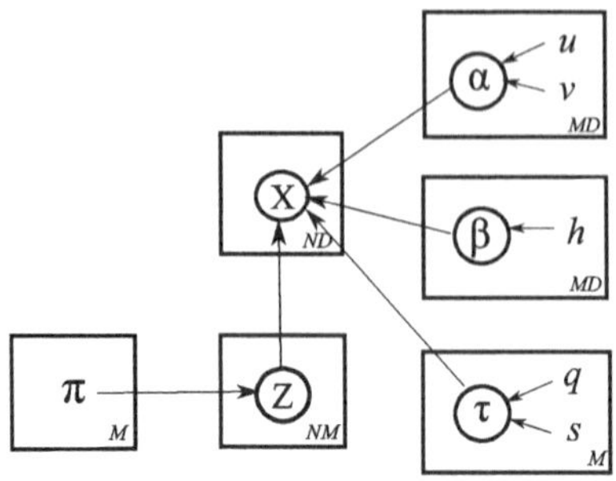

Figure 7.1 Graphical demonstration of the finite shifted-scaled Dirichlet mixture model.

By substituting Eqs. (7.10), (7.11), (7.12), (7.4) and (7.5) into the joint distribution defined in Eq. (7.6), we get

$$p(\mathcal{X}, \Phi \mid \vec{\pi}) = \prod_{i=1}^{N}\prod_{j=1}^{M}\Big(\pi_j \frac{\Gamma(\alpha_j+)}{\prod_{d=1}^{D}\Gamma(\alpha_{jd})}\frac{1}{\tau_j^{D-1}}\frac{\prod_{d=1}^{D}\beta_{jd}^{-\frac{\alpha_{jd}}{\tau_j}}x_{id}^{(\frac{\alpha_{jd}}{\tau_j}-1)}}{\left(\sum_{d=1}^{D}(\frac{x_{id}}{\beta_{jd}})^{\frac{1}{\tau_j}}\right)^{\alpha_j+}}\Big)^{Z_{ij}} \tag{7.13}$$

$$\times \prod_{j=1}^{M}\prod_{l=1}^{D}\Big[\frac{v_{jd}^{u_{jd}}}{\Gamma(u_{jd})}\alpha_{jd}^{u_{jd}-1}e^{-v_{jd}\alpha_{jd}} \times \frac{\Gamma(\sum_{d=1}^{D}h_{jd})}{\prod_{d=1}^{D}\Gamma(h_{jd})}\prod_{d=1}^{D}\beta_{jd}^{h_{jd}-1}\Big]$$

$$\times \prod_{j=1}^{M}\Big[\frac{q_j^{s_j}}{\Gamma(q_j)}\tau_j^{q_j-1}e^{-s_j\tau_j}\Big].$$

A graphical representation of this model is shown in Fig. 7.1 where random variables are displayed within circles. Plates denote replication and the number of replications is shown in the lower right corner of it. The arrows represent the conditional dependencies among variables.

For learning our mixture model parameters and defining the correct number of components M simultaneously, here we apply the variational inference approach proposed in [22]. In this technique, a tractable lower bound $\mathcal{L}(Q)$ on the marginal likelihood $p(\mathcal{X} \mid \vec{\pi})$ is used as below:

$$\ln p(\mathcal{X} \mid \vec{\pi}) = \ln \int p(\mathcal{X}, \Theta \mid \vec{\pi})d\Theta = \tag{7.14}$$

$$\ln \int Q(\Theta)\frac{p(\mathcal{X}, \Theta \mid \vec{\pi})}{Q(\Theta)}d\Theta \geq \int Q(\Theta)\ln\left(\frac{p(\mathcal{X}, \Theta \mid \vec{\pi})}{Q(\Theta)}\right)d\Theta = \mathcal{L}(Q)$$

where $Q(\Theta)$ is introduced as an approximation for the true posterior $p(\Theta \mid \mathcal{X}, \vec{\pi})$.

From Eq. (7.14), we find the following equation:

$$\mathcal{L}(Q) = \ln p(\mathcal{X} \mid \vec{\pi}) - KL(Q \,||\, P) \tag{7.15}$$

where

$$KL(Q \,||\, P) = -\int Q(\Theta) \ln\left(\frac{p(\Theta \mid \mathcal{X}, \vec{\pi})}{Q(\Theta)}\right) d\Theta. \tag{7.16}$$

By maximizing the lower bound $\mathcal{L}(Q)$, the KL divergence reaches its minimum, zero, when $Q(\Theta) = p(\Theta \mid \mathcal{X}, \vec{\pi})$. However, it is difficult to directly compute the true posterior for variational inference. Thus, $Q(\Theta)$ as a restricted family of distributions is taken into account.

We adopt an approximation method called mean-field theory [31], [32] which factorizes $Q(\Theta)$ into tractable distributions of each parameter in the parameter space Θ:

$$Q(\Theta) = Q(\mathcal{Z})Q(\vec{\alpha})Q(\vec{\beta})Q(\vec{\tau}). \tag{7.17}$$

Lower bound maximization is done through variational optimization of $\mathcal{L}(Q)$ with respect to each of the parameter distributions $Q_s(\Theta_s)$ which results in the following equation for a particular $Q_s(\Theta_s)$ [22]:

$$Q_s(\Theta_s) = \frac{exp\langle \ln p(\mathcal{X}, \Theta)\rangle_{j \neq s}}{\int exp\langle \ln p(\mathcal{X}, \Theta)\rangle_{j \neq s} d\Theta} \tag{7.18}$$

where $\langle \cdot \rangle_{j \neq s}$ indicates the expectation of all the distributions $Q_j(\Theta_j)$ excluding $j = s$. To apply the variational inference, all the parameter distributions $Q_j(\Theta_j)$ need to be initialized properly, since for optimal solution estimation $Q_s(\Theta_s)$ we loop over all the $Q_j(\Theta_j)$ except $j = s$. Afterward, each parameter gets updated by an improved value which is calculated from Eq. (7.18) assuming the recent value of the other parameters altogether. Due to the convexity of the lower bound corresponding to each of the parameter distributions $Q_j(\Theta_j)$, convergence is certain [33], [34]. Finally, we get the optimal variational estimations for each Q in the parameter space Θ as follows (see Appendix 7.A):

$$Q(\mathcal{Z}) = \prod_{i=1}^{N} \prod_{j=1}^{M} r_{ij}^{Z_{ij}}, \tag{7.19}$$

$$Q(\vec{\alpha}) = \prod_{j=1}^{M}\prod_{l=1}^{D}\mathcal{G}\left(\alpha_{jd} \mid u_{jd}^{*}, \nu_{jd}^{*}\right), \tag{7.20}$$

$$Q(\vec{\beta}) = \prod_{j=1}^{M}\prod_{l=1}^{D}\mathcal{D}\left(\beta_{jd} \mid h_{jd}^{*}\right), \tag{7.21}$$

$$Q(\vec{\tau}) = \prod_{j=1}^{M}\prod_{l=1}^{D}\mathcal{G}\left(\tau_{j} \mid q_{j}^{*}, s_{j}^{*}\right), \tag{7.22}$$

where

$$r_{ij} = \frac{\rho_{ij}}{\sum_{j=1}^{M}\rho_{ij}}, \tag{7.23}$$

$$\begin{aligned}\rho_{ij} = exp\Bigg\{ & \ln\pi_j + \tilde{R}_j - (D-1)\ln\overline{\tau}_j \\ & + \sum_{d=1}^{D}\left[-\frac{\overline{\alpha}_{jd}}{\overline{\tau}_j}\ln\overline{\beta}_{jd} + \left(\frac{\overline{\alpha}_{jd}}{\overline{\tau}_j} - 1\right)\ln x_{id}\right] \\ & - (\alpha_j +)\ln\left(\sum_{d=1}^{D}\left(\frac{x_{id}}{\overline{\beta}_{jd}}\right)^{\frac{1}{\overline{\tau}_j}}\right)\Bigg\}.\end{aligned} \tag{7.24}$$

$\tilde{R}_j$ is estimated as follows in which $\psi(.)$ and $\psi'(.)$ denotes digamma and trigamma functions, respectively (see Appendix 7.A):

$$\begin{aligned}\tilde{R}_j = & \ln\frac{\Gamma\left(\sum_{d=1}^{D}\overline{\alpha}_{jd}\right)}{\prod_{d=1}^{D}\Gamma\left(\overline{\alpha}_{jd}\right)} \\ & + \sum_{d=1}^{D}\overline{\alpha}_{jd}\left[\psi\left(\sum_{d=1}^{D}\overline{\alpha}_{jd}\right) - \psi\left(\overline{\alpha}_{jd}\right)\right] \\ & \times\left[\left\langle\ln\alpha_{jd}\right\rangle - \ln\overline{\alpha}_{jd}\right] \\ & + \frac{1}{2}\sum_{d=1}^{D}\overline{\alpha}_{jd}^{2}\left[\psi'\left(\sum_{d=1}^{D}\overline{\alpha}_{jd}\right) - \psi'\left(\overline{\alpha}_{jd}\right)\right] \\ & \times\left\langle\left(\ln\alpha_{jd} - \ln\overline{\alpha}_{jd}\right)^{2}\right\rangle \\ & + \frac{1}{2}\sum_{a=1}^{D}\sum_{b=1, a\neq b}^{D}\overline{\alpha}_{ja}\overline{\alpha}_{jb}\Bigg\{\psi'\left(\sum_{d=1}^{D}\overline{\alpha}_{jd}\right)\end{aligned} \tag{7.25}$$

$$\times\left(\left\langle\ln\overline{\alpha}_{ja}\right\rangle-\ln\overline{\alpha}_{ja}\right)\times\left(\left\langle\ln\overline{\alpha}_{jb}\right\rangle-\ln\overline{\alpha}_{jb}\right)\Bigg\}.$$

The hyper-parameters u_{jd}^*, v_{jd}^*, h_{jd}^*, q_j^* and s_j^* are approximated as below (see Appendix 7.A):

$$u_{jd}^*=u_{jd}+\varphi_{jd} \qquad v_{jd}^*=v_{jd}+\vartheta_{jd}, \tag{7.26}$$

$$\varphi_{jd}=\sum_{i=1}^{N}\langle Z_{ij}\rangle\overline{\alpha}_{jd}\left[\psi\left(\sum_{d=1}^{D}\overline{\alpha}_{jd}\right)-\psi\left(\overline{\alpha}_{jd}\right)+\sum_{d\neq s}^{D}\psi'\left(\sum_{d=1}^{D}\overline{\alpha}_{jd}\right)\right. \tag{7.27}$$
$$\left.\times\,\overline{\alpha}_{js}\left(\left\langle\ln\alpha_{js}\right\rangle-\ln\overline{\alpha}_{js}\right)\right],$$

$$\vartheta_{jd}=\sum_{i=1}^{N}\langle Z_{ij}\rangle\left[\frac{1}{\tau_j}\ln\frac{\beta_{js}}{x_{is}}+\ln\left(\sum_{d=1}^{D}\left(\frac{x_{id}}{\beta_{jd}}\right)^{\frac{1}{\tau_j}}\right)\right], \tag{7.28}$$

$$h_{jd}^*=h_{jd}+\kappa_{jd}, \tag{7.29}$$

$$\kappa_{jd}=\sum_{i=1}^{N}\langle Z_{ij}\rangle\left[\frac{-\bar{\alpha}_{js}}{\bar{\tau}_j}+\frac{\bar{\alpha}_{js}}{\bar{\tau}_j}\times\left(\frac{x_{is}}{\bar{\beta}_{js}}\right)^{\frac{1}{\bar{\tau}_j}}\times\frac{1}{\sum_{d=1}^{D}\left(\frac{x_{id}}{\bar{\beta}_{jd}}\right)^{\frac{1}{\bar{\tau}_j}}}\right], \tag{7.30}$$

$$q_j^*=q_j+\delta_j, \quad s_j^*=s_j-\varrho_j, \tag{7.31}$$

$$\delta_j=\sum_{i=1}^{N}Z_{ij}\left[1-D+\frac{(\alpha_j+)}{\tau_j}\frac{\sum_{d=1}^{D}\left(\frac{x_{id}}{\beta_{jd}}\right)^{\frac{1}{\tau_j}}\ln\left(\frac{x_{id}}{\beta_{jd}}\right)}{\sum_{d=1}^{D}\left(\frac{x_{id}}{\beta_{jd}}\right)^{\frac{1}{\tau_j}}}\right], \tag{7.32}$$

$$\varrho_j=\sum_{i=1}^{N}Z_{ij}\left[\sum_{d=1}^{D}\frac{\alpha_{jd}}{\tau_j^{2}}\ln\left(\frac{x_{id}}{\beta_{jd}}\right)\right], \tag{7.33}$$

where the expected values in the preceding equations are as follows:

$$\langle Z_{ij}\rangle=r_{ij}, \tag{7.34}$$

$$\overline{\alpha}_{jd}=\langle\alpha_{jd}\rangle=\frac{u_{jd}^*}{v_{jd}^*}, \qquad \langle\ln\alpha_{jd}\rangle=\psi\left(u_{jd}^*\right)-\ln v_{jd}^*, \tag{7.35}$$

$$\left\langle \left(\ln \alpha_{jd} - \ln \overline{\alpha}_{jd} \right)^2 \right\rangle = \left[\psi\left(u_{jd}^*\right) - \ln u_{jd}^* \right]^2 + \psi'\left(u_{jd}^*\right), \tag{7.36}$$

$$\overline{\beta}_{jd} = \left\langle \beta_{jd} \right\rangle = \frac{h_{jd}^*}{\sum_{d=1}^{D} h_{jd}^*}, \tag{7.37}$$

$$\overline{\tau}_j = \left\langle \tau_j \right\rangle = \frac{q_j^*}{s_j^*}. \tag{7.38}$$

7.3.2 Determination of the number of components

As mentioned before, $\vec{\pi}$ is considered as a parameter in variational learning. Thus, it is estimated via maximization of lower bond $\mathcal{L}(Q)$. The equation is obtained via setting the derivative of the lower bound with respect to $\vec{\pi}$ to zero:

$$\pi_j = \frac{1}{N} \sum_{i=1}^{N} r_{ij}. \tag{7.39}$$

As the lower bound $\mathcal{L}(Q)$ is maximized in order to obtain the variational optimization of $Q(\mathcal{Z})$, $Q(\vec{\alpha})$, $Q(\vec{\beta})$, and $Q(\vec{\tau})$, the mixing coefficient $\vec{\pi}$ gets estimated as well. Therefore, the components which have trivial contribution to describe the data would have a close-to-zero mixing coefficients. Using automatic relevance determination [35], these components would be omitted from the model.

The steps of the variational algorithm for our model are described in Algorithm 1.

7.4 Experimental result

In this section, we evaluate the performance of our proposed model using three medical datasets, namely, malaria, breast cancer and heart diseases as well as a challenging text dataset, namely spam detection. The accuracy of the model mainly relies on the initialization of the hyper-parameters including $\{u_{jd}\}$, $\{v_{jd}\}$, $\{h_{jd}\}$, $\{q_{jd}\}$, and $\{s_{jd}\}$. Thus, detecting a good set of initialized hyper-parameters is an important step to obtaining the optimal number of clusters and enhance the convergence rate. Besides, feature extraction and feature selection techniques are inevitably part of the data pre-processing approach when dealing with image and text applications. Furthermore, scaling and normalization are crucial steps and need to be always considered for total performance improvement. To enhance the outperformance of our model, we compare it with four other

Algorithm 1 Variational learning algorithm of SSD.

1. Initialize number of components M.
2. Initialize the hyper-parameters $\{u_{jd}\}$, $\{v_{jd}\}$, $\{h_{jd}\}$, $\{q_{jd}\}$ and $\{s_{jd}\}$.
3. Initialize r_{ij} using *K-Means* algorithm.
4. **repeat**
5. E-step of variational: update $Q(\mathcal{Z})$, $Q(\vec{\alpha})$, $Q(\vec{\beta})$ and $Q(\vec{\tau})$.
6. M-step of variational: Maximize $\mathcal{L}(Q)$ with respect to recent value of $\vec{\pi}$ (7.39).
7. **until** Convergence criterion is reached.
8. Determine the number of components M by omitting those with trivial mixing coefficients (smaller than 10^{-5}).
9. Re-estimate new values of the parameters $(\mathcal{Z})$, $(\vec{\alpha})$, $(\vec{\beta})$, $(\vec{\tau})$, and $(\vec{\pi})$.

models, namely, variational learning of a scaled Dirichlet mixture model (varSDMM), variational learning of a Dirichlet mixture model (varDMM), variational learning of a Gaussian mixture model (varGMM) and maximum likelihood learning of a Gaussian mixture model (GMM).

7.4.1 Malaria detection

Malaria is a fatal disease in countries with tropical climates. It is caused by a parasite which is transmitted to humans through the bite of an infected mosquito. Based on the latest report released by WHO, in 2018, 405000 malaria deaths have been registered among 228 million cases worldwide [36]. An accurate diagnosis is crucial in order to prevent from death and prevalence. Parasitological and clinical microscopy as a commonly used mean involves visual analysis of blood smears in order to detect the parasite in the blood as well as identifying the type, number and life cycle of the parasitemia. However, microscopy examination could be overwhelming and costly and very much relies on the qualification of the specialist and load of samples. The need for sample analysis automation has become undeniable considering the recent report published by WHO that 207 million suspected patients were tested via either an RDT or microscopy in 2018 [36]. We obtained a dataset from NIH containing slide images of blood

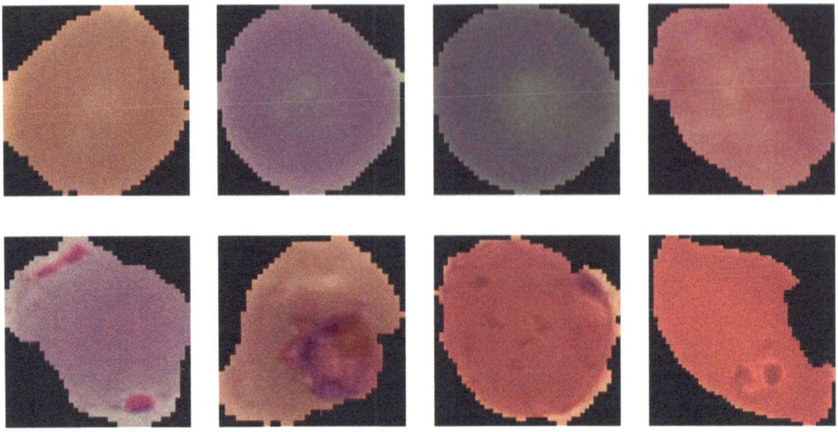

Figure 7.2 Malaria dataset.

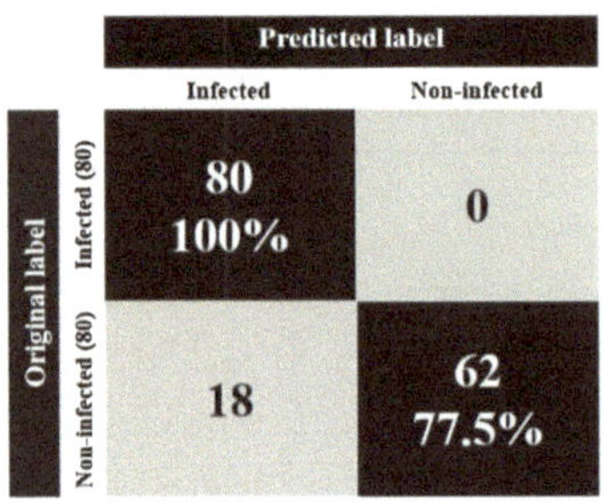

Figure 7.3 Malaria confusion matrix.

smear released by the malaria screener research activity [37]. The dataset has 27,558 images of blood cells with equal instances of infected and normal ones. Some samples of this dataset are shown in Fig. 7.2, containing infected and normal blood smear instances. Acquiring a precise representation of the features of a dataset is an essential pre-processing task. In other words, an efficient descriptor containing most of the important features is needed. For this dataset we used Bag of visual words (BOVW) and SIFT [38] since it has well performed in various classification problems [33], [39], [40]. From the confusion matrix in Fig. 7.3, we can see that 100% of the infected cells and 77.5% of the non-infected ones have been accurately detected which we can compare with other algorithms' confusion matrices shown in Fig. 7.4. Finally, we compared the result of our model with four other models summarized in Table 7.1 denoting the outstanding accu-

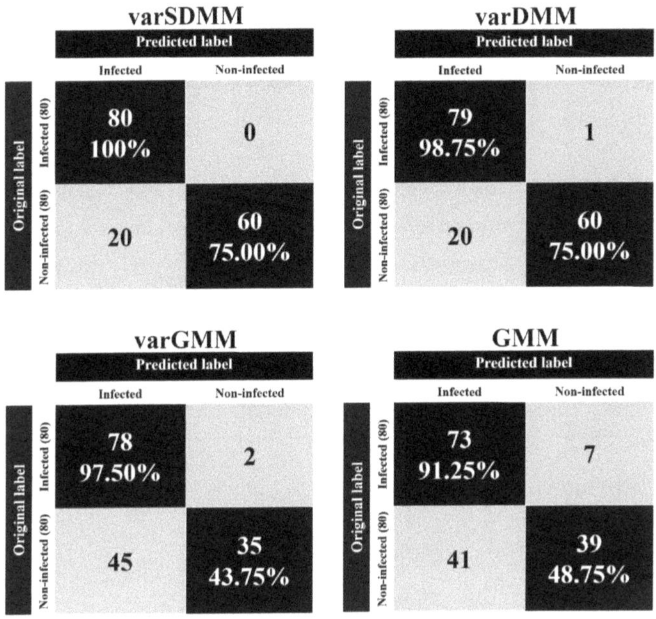

Figure 7.4 Malaria confusion matrices for other algorithms.

Table 7.1 Model performance accuracy in malaria dataset.

Algorithm	varSSDMM	varSDMM	varDMM	varGMM	GMM
Accuracy(%)	88.7	87.5	86.8	70.6	70.0

racy of varSSDMM (88.7%). These results endorse the variational learning method on the shifted-scaled Dirichlet technique as an effective approach.

7.4.2 Breast cancer diagnosis

Breast cancer is the most prevalent type of cancers in women and the second common cancer worldwide, according to WCRF [41]. Early diagnosis and treatment play key roles in improving the survival rate of cancerous patients. Most of the breast cancer detections are done via screening imaging like sonography, mammography or MRI. Once a lump is detected, it is sampled for further analysis. Then a pathologist examines the tissue sample for detection of being benign or malignant. Among different ways of sampling, the fine needle aspiration (FNA) [42] is one of the standard and suitable means for medical diagnostic and decision-making processes. We evaluate our model over a publicly available breast cancer dataset named

Attribute number	Attribute description	Values of attributes	Mean	SD
1	Clump thickness	1–10	4.42	2.82
2	Uniformity of cell size	1–10	3.13	3.05
3	Uniformity of cell shape	1–10	3.20	2.97
4	Marginal adhesion	1–10	2.80	2.86
5	Single epithelial cell size	1–10	3.21	2.21
6	Bare nuclei	1–10	3.46	3.64
7	Bland chromatin	1–10	3.43	2.44
8	Normal nucleoli	1–10	2.87	3.05
9	Mitoses	1–10	1.59	1.71

Figure 7.5 Attributes of Wisconsin Dataset.

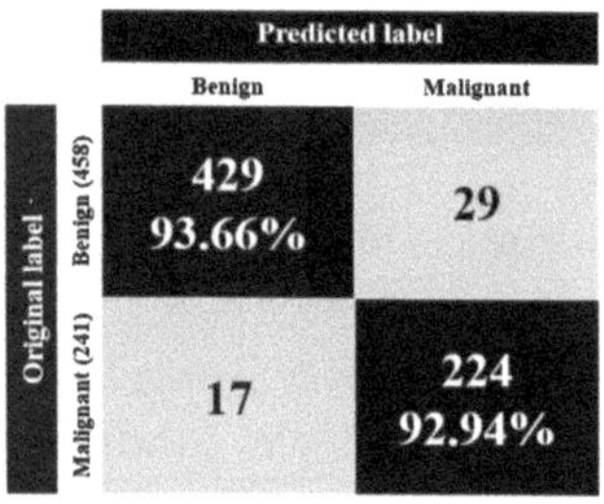

Figure 7.6 Wisconsin confusion matrix.

Wisconsin [43]. This dataset contains 699 samples with nine features that are computed from the images of breast lumps obtained via fine needle aspirate. In Fig. 7.5, the attributes are graded from 1 to 10 with 1 being the closest to benign and 10 the most anaplastic and malignant [44]. The mean and standard deviation of each attribute is detailed as well. It is noteworthy to mention that no single feature alone is enough to differentiate among benign and malignant instances and we employ all of them. There are (458) benign and (241) malignant cases in the set. The confusion matrix presented in Fig. 7.6 clearly shows that the majority of the instances are correctly categorized, compared to other algorithms' confusion matrices shown in Fig. 7.7. Lastly, the final result in Table 7.2 shows a significant improvement in clustering the benign and malignant samples using varSS-DMM (93.4% accuracy) compared to the rest of algorithms. This result is

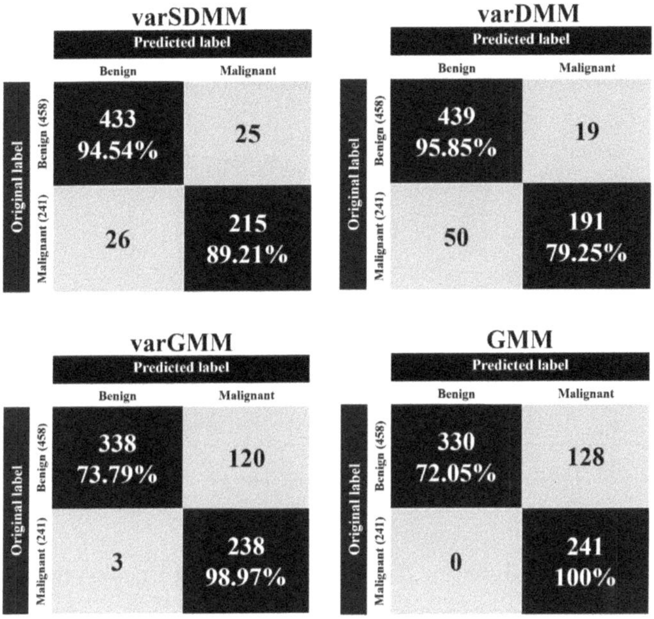

Figure 7.7 Wisconsin confusion matrices for other algorithms.

Table 7.2 Model performance accuracy in Wisconsin dataset.

Algorithm	varSSDMM	varSDMM	varDMM	varGMM	GMM
Accuracy(%)	93.4	92.7	90.1	82.7	81.6

achieved given that min-max scaling and normalization is applied on the dataset in order to improve the final result.

7.4.3 Cardiovascular diseases (CVDs) detection

Cardiovascular diseases (CVDs) as a wide assortment of disorders influencing the heart and veins are perceived as the first reason for worldwide death. This leading explanation of mortality has ended the lives of 17.9 million individuals each year [45]. It is overwhelmingly costly to diagnose and control these illnesses due to the need for long-term treatment and pricey equipment. Thus, CVDs carries loads of expenses imposed to medicinal services and consequently government. However, considering the related risk factors of heart diseases such as obesity, tobacco use, low physical movement and diet, prevention could always be an essential approach. These days, complex data such as clinical history, biomarkers, pictures, signals and text

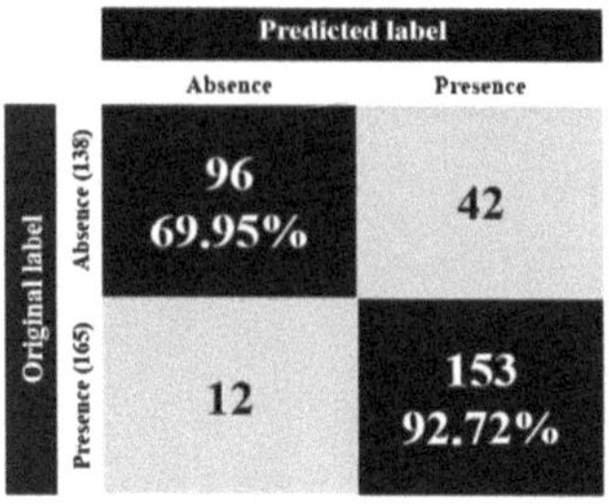

Figure 7.8 Heart disease confusion matrix.

Table 7.3 Model performance accuracy in heart dataset.

Algorithm	**varSSDMM**	**varSDMM**	**varDMM**	**varGMM**	**GMM**
Accuracy(%)	82.2	80.1	77.5	71.9	71.3

are the source of analysis for doctors which could be a complicated task. Therefore, such a diagnosis system could be error-prone, inaccurate and could put the patient in danger. In these situations, automation in clinical inference could be helpful [46]. In this part of our experiment, we evaluated our proposed model over a real and publicly available dataset [47] to predict heart disease existence based on specific characteristics of a person. There are 303 samples with 76 attributes being measured in this dataset, but all released experiments have utilized a subset of 14 features including age, sex, chest pain location, resting blood pressure, serum cholesterol level, fasting blood sugar, resting electrocardiographic results (normal, ST-T wave abnormality or left ventricular hypertrophy), maximum heart rate achieved, exercise induced angina, ST depression peak, the slope of the peak exercise ST segment (upsloping, flat or downsloping), number of major vessels colored by fluoroscopy and type of defect (normal, fixed or reversible). The confusion matrix in Fig. 7.8 confirms that most of the instances have been well classified, especially 92% of the heart disease presences has been detected which we can compare with other algorithms' confusion matrices in Fig. 7.9. The outcome of our assessment is demonstrated in Table 7.3 denoting the outperformance of varSSDMM with 82% overall accuracy. It is noteworthy to mention that reprocessing techniques such as min-max scaling and normalization have been performed on the dataset before applying our model.

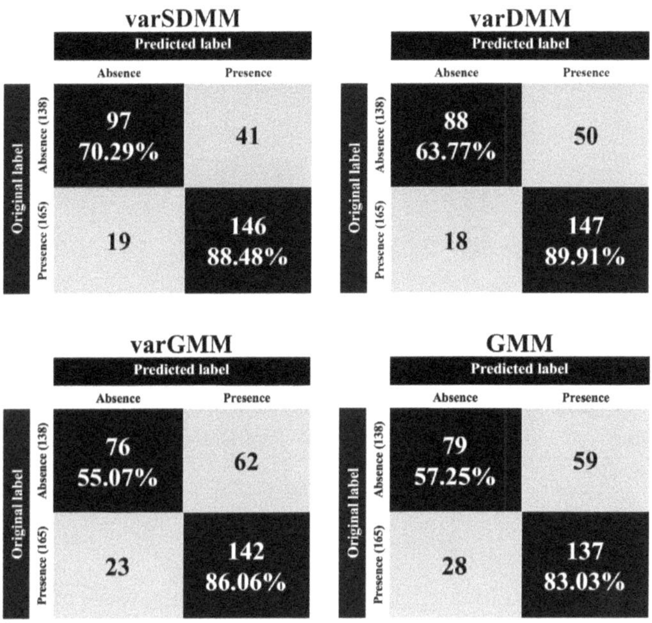

Figure 7.9 Heart disease confusion matrices for other algorithms.

7.4.4 Spam detection

The fourth real application we have tested our model on is a text application. One of the serious research fields in information system security is spam detection. The concept of spam or unsolicited message is extended from product or website advertisements, money-making scams, pornography to chain letters. The most widely recognized form of spam is email spam which creates major problems such as lost productivity, financial damage and fraud. According to some references around 80% of emails are spam which brought about overwhelming financial losses of 50 billion dollars in 2005 [48]. Among all of the methodologies developed to stop spam, filtering is a significant and mainstream one. Applying machine learning and pattern recognition methods have significantly improved spam filtering compare to other user-defined rules [49], [50]. For our experiment, we obtained a challenging spam dataset from UCI machine learning repository, provided by Hewlett-Packard Labs [51]. The dataset has 4601 instances with 57 continuous input attributes plus the target column which denotes whether the e-mail was categorized spam (1) or not (0). Among all the instances, 39.4% of them (1813) are spam and 60.6% (2788) are non-spam.

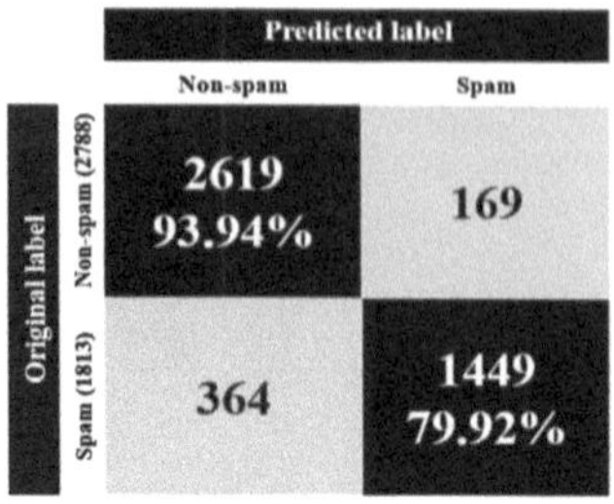

Figure 7.10 Spam detection confusion matrix.

Table 7.4 Model performance accuracy in spam detection dataset.

Algorithm	varSSDMM	varSDMM	varDMM	varGMM	GMM
Accuracy(%)	88.3	79.5	77.2	75.0	74.7

These attributes are obtained through a commonly used method called Bag of Words (BoW) [52] which is one of the effective data representation techniques in natural language processing. A majority of the attributes indicate the frequency of a particular word or character appearance in e-mail. In other words, each email is represented by its words ignoring grammar. Forty-eight attributes contain the percentage of the respective word and 6 attributes include the percentage of the respective characters in the e-mail. The rest of them are the average length of continuous sequences of capital letters, the length of the longest continuous sequence of capital letters and the total number of capital letters in the e-mail. We carried out some pre-processing steps on the dataset before applying our model in order to enhance the final result. These steps include employing feature selection techniques which reduced the number of features to 45. Afterwards, min-max scaling and normalization is performed to further enhance our accuracy. In Fig. 7.10, the outcome of our performance, represented in the confusion matrix, denoting better classification of most of the e-mails compared to other algorithms' confusion matrices in Fig. 7.11. The summary of our model accuracy compared to four other algorithms represented in Table 7.4 that with 88.3% accuracy confirms the advantage of the model.

Thus, we have evaluated our model on four different sizes of datasets and compared the results with three other variational models (varSDMM,

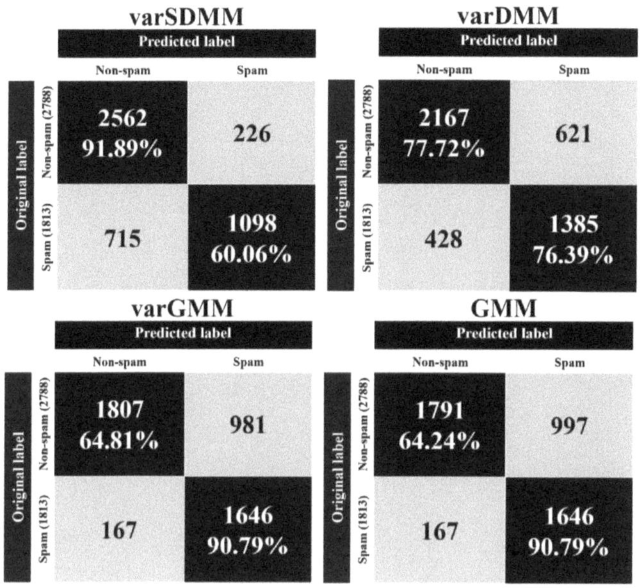

Figure 7.11 Spam detection confusion matrices for other algorithms.

varDMM, varGMM) as well as a deterministic model (GMM) to prove the potency and robustness of this model.

7.5 Conclusion

We have effectively applied variational Bayesian learning method on a finite shifted-scaled Dirichlet mixture model. Using this model, parameters' approximation was precisely accomplished avoiding the heavy cost of computation associated with conventional Bayesian strategies. Besides, the number of clusters was simultaneously detected as part of a lower-bound maximization procedure. Our model has proven to be more effective than two other similar models in medical real applications, including malaria detection, breast cancer diagnosis and cardiovascular diseases as well as a challenging text application like spam detection. Further enhancement of the model could include adding feature selection, upgrading to an infinite model or developing online learning framework.

Appendix 7.A

In this section, we present the proofs for (7.19), (7.20), (7.21), and (7.22).

According to (7.18), we can rewrite the general expression of the variational solution $Q_s(\Theta_s)$ as

$$\ln Q_s(\Theta_s) = \langle \ln p(\mathcal{X}, \Theta) \rangle_{j \neq s} + const, \tag{7.40}$$

in which those terms that are independent of the respective parameter in $Q_s(\Theta_s)$, are assimilated into the constant. Utilizing Eq. (7.40) along with the logarithm of the joint distribution in (7.13), $p(\mathcal{X}, \Phi \mid \vec{\pi})$, we calculated variational solutions for each parameter as follows.

A. Proof of (7.19): Variational solution to $Q(\mathcal{Z})$

$$\begin{aligned}\ln Q(\mathcal{Z}_{ij}) = \mathcal{Z}_{ij} \Bigg\{ & \ln \pi_j + R_j - (D-1) \ln \overline{\tau}_j \\ & + \sum_{d=1}^{D} \left[-\frac{\overline{\alpha}_{jd}}{\overline{\tau}_j} \ln \overline{\beta}_{jd} + (\frac{\overline{\alpha}_{jd}}{\overline{\tau}_j} - 1) \ln x_{id} \right] \\ & - (\alpha_j +) \ln \left(\sum_{d=1}^{D} (\frac{x_{id}}{\overline{\beta}_{jd}})^{\frac{1}{\overline{\tau}_j}} \right) \Bigg\} + const\end{aligned} \tag{7.41}$$

where $R_j = \left\langle \ln \frac{\Gamma\left(\sum_{d=1}^{D} \alpha_{jd}\right)}{\prod_{d=1}^{D} \Gamma\left(\alpha_{jd}\right)} \right\rangle_{\alpha_{j1}, \ldots, \alpha_{jD}}$,

$$\overline{\alpha}_{jd} = \langle \alpha_{jd} \rangle = \frac{u_{jd}^*}{v_{jd}^*}, \tag{7.42}$$

However, R_j is analytically intractable and we are not able to directly perform variational inference. Thus, we need to approximate a lower bound for it that gives us a closed-form expression. Obtaining a tractable approximation with applying second order Taylor series expansion in variational inference has been effectively done in [53], [54]. Moreover, we can find the same function R_j approximated using a second order Taylor series in [33] that we will utilize here. The approximation of R_j around the expected values of $\vec{\alpha}_j$, represented by $(\overline{\alpha}_{j1}, \ldots, \overline{\alpha}_{jD})$ is defined as $\tilde{R}_j$ and denoted in (7.25). Now the equation in (7.41) turns into a tractable expression after substituting R_j by $\tilde{R}_j$ and we can obviously notice that the optimal solution estimation to $\mathcal{Z}$ takes the logarithmic form of (7.4) excluding the normalization constant. Therefore, we can rewrite $\ln Q(\mathcal{Z})$ as follows:

$$\ln Q(\mathcal{Z}) = \sum_{i=1}^{N}\sum_{j=1}^{M} \mathcal{Z}_{ij} \ln \rho_{ij} + const, \tag{7.43}$$

$$\begin{aligned} \ln \rho_{ij} = {} & \ln \pi_j + \tilde{R}_j - (D-1)\ln \overline{\tau}_j \\ & + \sum_{d=1}^{D}\left[-\frac{\overline{\alpha}_{jd}}{\overline{\tau}_j} \ln \overline{\beta}_{jd} + (\frac{\overline{\alpha}_{jd}}{\overline{\tau}_j} - 1)\ln x_{id} \right] \\ & - (\alpha_j +) \ln\left(\sum_{d=1}^{D} (\frac{x_{id}}{\overline{\beta}_{jd}})^{\frac{1}{\overline{\tau}_j}} \right). \end{aligned} \tag{7.44}$$

By taking the exponential of both sides in (7.43), we get

$$Q(\mathcal{Z}) \propto \prod_{i=1}^{N}\prod_{j=1}^{M} \rho_{ij}^{Z_{ij}}; \tag{7.45}$$

after applying normalization on the previous distribution, we obtain

$$Q(\mathcal{Z}) = \prod_{i=1}^{N}\prod_{j=1}^{M} r_{ij}^{Z_{ij}}, \qquad r_{ij} = \frac{\rho_{ij}}{\sum_{j=1}^{M} \rho_{ij}}, \tag{7.46}$$

note that the $\{r_{ij}\}$ are non-negative and sum to one. Thus, the standard solution for $Q(\mathcal{Z})$ can be derived as

$$\langle Z_{ij} \rangle = r_{ij} \tag{7.47}$$

where $\{r_{ij}\}$ are equivalent to responsibilities in the conventional EM algorithms.

B. Proof of (7.20): Variational solution to $Q(\vec{\alpha})$

Since the parameters are considered statistically independent and there are M clusters in the mixture model, $Q(\vec{\alpha})$ can be factorized as follows:

$$Q(\vec{\alpha}) = \prod_{j=1}^{M}\prod_{d=1}^{D} Q(\alpha_{jd}). \tag{7.48}$$

The logarithm of the optimized factor with respect to the specific parameter α_{js} is calculated as follows:

$$\ln Q(\alpha_{js}) = \sum_{i=1}^{N} r_{ij}\mathcal{J}(\alpha_{js}) - \frac{\alpha_{js}}{\tau_j}\ln\beta_{js}\sum_{i=1}^{N} r_{ij} + \frac{\alpha_{js}}{\tau_j}\sum_{i=1}^{N} r_{ij}\ln x_{is} - \alpha_{js}\sum_{i=1}^{N} r_{ij}\ln\left(\sum_{d=1}^{D}\left(\frac{x_{id}}{\beta_{jd}}\right)^{\frac{1}{\tau_j}}\right) + (u_{js}-1)\ln\alpha_{js} - \nu_{js}\alpha_{js} + const \tag{7.49}$$

where $\mathcal{J}(\alpha_{js}) = \left\langle \ln\frac{\Gamma(\alpha_s + \sum_{d\neq s}^{D}\alpha_{jd})}{\Gamma(\alpha_s)\prod_{d\neq s}^{D}\Gamma(\alpha_{jd})}\right\rangle_{\Theta\neq\alpha_{js}}$ is described as a function of $\alpha_{\alpha_{js}}$, which unfortunately does not have a closed-form solution. Therefore, as for R_j in the section A, we need to approximate $\mathcal{J}(\alpha_{js})$ by finding a lower bound via a Taylor series expansion about $\overline{\alpha}_{js}$ (the expected value of α_{js}). The same function has been approximated in [33] (Appendix 7.B) and we shall use the final result here:

$$\mathcal{J}(\alpha_{js}) \geq \overline{\alpha}_{js}\ln\alpha_{js}\left\{\Psi\left(\sum_{d=1}^{D}\overline{\alpha}_{jd}\right) - \Psi(\overline{\alpha}_{js}) + \sum_{d\neq s}^{D}\overline{\alpha}_{jd} \times \Psi'\left(\sum_{d=1}^{D}\overline{\alpha}_{jd}\right)(\langle\ln\alpha_{jd}\rangle - \ln\overline{\alpha}_{jd})\right\} + const; \tag{7.50}$$

after substituting this lower bound back into (7.49), we get a new optimal solution to α_{js} as follows:

$$\begin{aligned}\ln Q(\alpha_{js}) &= \sum_{i=1}^{N} r_{ij}\overline{\alpha}_{js}\ln\alpha_{js}\left[\Psi\left(\sum_{d=1}^{D}\overline{\alpha}_{jd}\right) - \Psi(\overline{\alpha}_{js}) + \sum_{d\neq s}^{D}\Psi'\left(\sum_{d=1}^{D}\overline{\alpha}_{jd}\right) \times \overline{\alpha}_{jd}(\langle\ln\alpha_{jd}\rangle - \ln\overline{\alpha}_{jd})\right] - \frac{\alpha_{js}}{\tau_j}\ln\beta_{js}\sum_{i=1}^{N} r_{ij} + \frac{\alpha_{js}}{\tau_j}\sum_{i=1}^{N} r_{ij}\ln x_{is} \\ &\quad - \alpha_{js}\sum_{i=1}^{N} r_{ij}\ln\left(\sum_{d=1}^{D}\left(\frac{x_{id}}{\beta_{jd}}\right)^{\frac{1}{\tau_j}}\right) + (u_{js}-1)\ln\alpha_{js} - \nu_{js}\alpha_{js} \\ &= \ln\alpha_{js}(u_{js} + \varphi_{js} - 1) - \alpha_{js}(\nu_{js} + \vartheta_{js}) + const\end{aligned} \tag{7.51}$$

where

$$\varphi_{jd} = \sum_{i=1}^{N} r_{ij}\overline{\alpha}_{jd}\left[\psi\left(\sum_{d=1}^{D}\overline{\alpha}_{jd}\right) - \psi(\overline{\alpha}_{jd}) + \sum_{d\neq s}^{D}\psi'\left(\sum_{d=1}^{D}\overline{\alpha}_{jd}\right)\right.$$
$$\left.\times\,\overline{\alpha}_{js}\left(\langle\ln\alpha_{js}\rangle - \ln\overline{\alpha}_{js}\right)\right], \tag{7.52}$$

$$\vartheta_{jd} = \sum_{i=1}^{N} r_{ij}\left[\frac{1}{\tau_j}\ln\frac{\beta_{js}}{x_{is}} + \ln\left(\sum_{d=1}^{D}\left(\frac{x_{id}}{\beta_{jd}}\right)^{\frac{1}{\tau_j}}\right)\right]. \tag{7.53}$$

We notice that (7.51) has gotten a logarithmic form of a Gamma function. If we take the exponential of both sides, we get

$$Q(\alpha_{js}) \propto \alpha_{js}^{u_{js}+\varphi_{js}-1}e^{-\alpha_{js}(\nu_{js}+\vartheta_{js})}. \tag{7.54}$$

Thus, we can derive the optimal solution to the hyper-parameters u_{js} and ν_{js} as

$$u_{js}^{*} = u_{js} + \varphi_{js}, \qquad \nu_{js}^{*} = \nu_{js} + \vartheta_{js}. \tag{7.55}$$

C. Proof of (7.21): Variational solution to $Q(\vec{\beta})$

Considering the assumption of parameter independence, for the M cluster in the mixture model, $Q(\vec{\beta})$ can be factorized as follows:

$$Q(\vec{\beta}) = \prod_{j=1}^{M}\prod_{d=1}^{D} Q(\beta_{jd}). \tag{7.56}$$

The logarithm of the optimized factor with respect to the specific parameter β_{js} is calculated as follows:

$$\ln Q(\beta_{js}) = -\frac{\alpha_{js}}{\tau_j}\ln\beta_{js}\sum_{i=1}^{N} Z_{ij} - (\alpha_j +)\sum_{i=1}^{N} Z_{ij}\mathcal{F}(\beta_{js})$$
$$+ (h_{js} - 1)\ln\beta_{js} + const \tag{7.57}$$

where

$$\mathcal{F}(\beta_{js}) = \left\langle \ln\left(\sum_{d=1}^{D}\left(\frac{x_{id}}{\beta_{jd}}\right)^{\frac{1}{\tau_j}}\right)\right\rangle. \tag{7.58}$$

We can notice that $\mathcal{F}(\beta_{js})$ is analytically intractable and needs to be approximated (see Appendix 7.B). The lower bound is approximated about $\overline{\beta}_{js}$, as follows:

$$\mathcal{F}(\beta_{js}) \geq (\frac{x_{is}}{\overline{\beta}_{js}})^{\frac{1}{\overline{\tau}_j}} \frac{-\ln \beta_{js}}{\overline{\tau}_j \sum_{d=1}^{D} (\frac{x_{id}}{\overline{\beta}_{jd}})^{\frac{1}{\overline{\tau}_j}}}. \tag{7.59}$$

By replacing this lower bound back into (7.57), we have the following equation:

$$\begin{aligned} \ln(\beta_{js}) = &-\frac{\alpha_{js}}{\tau_j} \ln \beta_{js} \sum_{i=1}^{N} r_{ij} - (\alpha_j +) \sum_{i=1}^{N} r_{ij} \\ &\times \left[(\frac{x_{is}}{\overline{\beta}_{js}})^{\frac{1}{\overline{\tau}_j}} \frac{-\ln \beta_{js}}{\overline{\tau}_j \sum_{d=1}^{D} (\frac{x_{id}}{\overline{\beta}_{jd}})^{\frac{1}{\overline{\tau}_j}}} \right] + (h_{js} - 1) \ln \beta_{js} + const, \\ = &(h_{js} + \kappa_{js} - 1) \ln \beta_{js} + const, \end{aligned} \tag{7.60}$$

where

$$\kappa_{js} = \sum_{i=1}^{N} r_{ij} \left[\frac{-\bar{\alpha}_{js}}{\bar{\tau}_j} + \frac{\bar{\alpha}_{js}}{\bar{\tau}_j} (\frac{x_{is}}{\bar{\beta}_{js}})^{\frac{1}{\bar{\tau}_j}} \frac{1}{\sum_{d=1}^{D} (\frac{x_{id}}{\bar{\beta}_{jd}})^{\frac{1}{\bar{\tau}_j}}} \right]. \tag{7.61}$$

We can notice that (7.60) has gotten a logarithmic form of a Beta distribution. By taking the exponential of both sides, we get

$$Q(\beta_{js}) \propto \beta_{js}^{h_{js} + \kappa_{js} - 1}. \tag{7.62}$$

Therefore, we can extract the optimal solution to the hyper-parameter h_{js} as

$$h_{js}^{*} = h_{js} + \kappa_{js}. \tag{7.63}$$

D. Proof of (7.22): Variational Solution to $Q(\vec{\tau})$

For an M cluster in the mixture model, we can factorize $Q(\vec{\tau})$ as follows:

$$Q(\vec{\tau}) = \prod_{j=1}^{M} Q(\tau_j). \tag{7.64}$$

By taking the logarithm of the optimized factor with respect to the specific parameter τ_j, we get

$$\begin{aligned}\ln Q(\tau_j) = {} & (1 - D)\ln \tau_j \sum_{i=1}^{N} r_{ij} + \sum_{i=1}^{N} r_{ij} \sum_{d=1}^{D} \frac{\alpha_{jd}}{\tau_j} (\ln x_{id} - \ln \beta_{jd}) \\ & - (\alpha_j +) \sum_{i=1}^{N} r_{ij} \mathcal{G}(\tau_j) + (q_j - 1)\ln \tau_j - s_j \tau_j + const\end{aligned} \tag{7.65}$$

where

$$\mathcal{G}(\tau_j) = \left\langle \ln \left(\sum_{d=1}^{D} \left(\frac{x_{id}}{\beta_{jd}}\right)^{\frac{1}{\tau_j}}\right) \right\rangle. \tag{7.66}$$

This is a function of τ_j, again analytically intractable; it needs to be approximated (see Appendix 7.B). We obtain the approximated lower bound, about $\overline{\tau}_j$, as follows:

$$\mathcal{G}(\tau_j) \geq \frac{-\ln \tau_j}{\overline{\tau}_j} \frac{\sum_{d=1}^{D} \left(\frac{x_{id}}{\overline{\beta}_{jd}}\right)^{\frac{1}{\overline{\tau}_j}} \ln \left(\frac{x_{id}}{\overline{\beta}_{jd}}\right)}{\sum_{d=1}^{D} \left(\frac{x_{id}}{\overline{\beta}_{jd}}\right)^{\frac{1}{\overline{\tau}_j}}} + const. \tag{7.67}$$

By substituting this lower bound back into (7.65), we have the following equation:

$$\begin{aligned}\ln Q(\tau_j) = {} & (1 - D)\ln \tau_j \sum_{i=1}^{N} r_{ij} + \sum_{i=1}^{N} r_{ij} \sum_{d=1}^{D} \frac{\alpha_{jd}}{\tau_j} \\ & \times (\ln x_{id} - \ln \beta_{jd}) + \frac{\ln \tau_j}{\overline{\tau}_j} (\alpha_j +) \sum_{i=1}^{N} r_{ij}\end{aligned} \tag{7.68}$$

$$
\times \frac{\sum_{d=1}^{D}(\frac{x_{id}}{\beta_{jd}})^{\frac{1}{\tau_j}} \ln(\frac{x_{id}}{\beta_{jd}})}{\sum_{d=1}^{D}(\frac{x_{id}}{\beta_{jd}})^{\frac{1}{\tau_j}}} + (q_j - 1)\ln\tau_j - s_j\tau_j + const,
$$
$$
= \ln\tau_j(q_j + \delta_j - 1) - \tau_j(s_j - \varrho_j),
$$

where

$$
\delta_j = \sum_{i=1}^{N} r_{ij}\left[1 - D + \frac{(\alpha_j +)}{\tau_j} \frac{\sum_{d=1}^{D}(\frac{x_{id}}{\beta_{jd}})^{\frac{1}{\tau_j}} \ln(\frac{x_{id}}{\beta_{jd}})}{\sum_{d=1}^{D}(\frac{x_{id}}{\beta_{jd}})^{\frac{1}{\tau_j}}}\right], \tag{7.69}
$$

$$
\varrho_j = \sum_{i=1}^{N} r_{ij}\left[\sum_{d=1}^{D} \frac{\alpha_{jd}}{\tau_j^2} \ln(\frac{x_{id}}{\beta_{jd}})\right]. \tag{7.70}
$$

We can see that (7.68) has gotten the logarithmic form of a Gamma function. By taking the exponential of both sides, we get

$$
Q(\tau_j) \propto \tau_j^{q_j+\delta_j-1} e^{-\tau_j(s_j-\varrho_j)}. \tag{7.71}
$$

So, we can obtain the optimal solution to the hyper-parameters q_j and s_j:

$$
q_j^* = q_j + \delta_j, \qquad s_j^* = s_j - \varrho_j. \tag{7.72}
$$

Appendix 7.B

A. Proof of (7.59): Lower bound of $\mathcal{F}(\beta_{js})$

Let us define the function $\mathcal{F}(\beta_{js})$ as

$$
\mathcal{F}(\beta_{js}) = \ln\left((\frac{x_{is}}{\beta_{js}})^{\frac{1}{\tau_j}} + \sum_{d \neq s}^{D}(\frac{x_{id}}{\beta_{jd}})^{\frac{1}{\tau_j}}\right). \tag{7.73}
$$

Since $\mathcal{F}(\beta_{js})$ is a convex function with respect to $\ln\beta_{js}$, we can calculate its lower bound using a first-order Taylor expansion of $\mathcal{F}(\beta_{js})$ for $\ln\beta_{js}$ at

$\ln \beta_{js,0}$ as below:

$$\mathcal{F}(\beta_{js}) \geq \mathcal{F}(\beta_{js,0}) + \frac{\partial \mathcal{F}(\beta_{js})}{\partial \ln \beta_{js}}|_{\beta_{js}=\beta_{js,0}}(\ln \beta_{js} - \ln \beta_{js,0}), \tag{7.74}$$

$$= \ln\left(\left(\frac{x_{is}}{\beta_{js,0}}\right)^{\frac{1}{\tau_j}} + \sum_{d\neq s}^{D}\left(\frac{x_{id}}{\beta_{jd,0}}\right)^{\frac{1}{\tau_j}}\right),$$

$$+ \beta_{js,0}\frac{1}{\left(\frac{x_{is}}{\beta_{js,0}}\right)^{\frac{1}{\tau_j}} + \sum_{d\neq s}^{D}\left(\frac{x_{id}}{\beta_{jd,0}}\right)^{\frac{1}{\tau_j}}}\left(\frac{-1}{\tau_j}X_{is}^{\frac{1}{\tau_j}}\beta_{js,0}^{\frac{-1}{\tau_j}-1}\right)(\ln \beta_{js} - \ln \beta_{js,0}),$$

$$= \left(\frac{x_{is}}{\beta_{js,0}}\right)^{\frac{1}{\tau_j}}\frac{-1}{\tau_j \sum_{d=1}^{D}\left(\frac{x_{id}}{\beta_{jd}}\right)^{\frac{1}{\tau_j}}} \times \ln \beta_{js}.$$

B. Proof of (7.67): Lower bound of $\mathcal{G}(\tau_j)$

Let us define the function $\mathcal{G}(\tau_j)$ as

$$\mathcal{G}(\tau_j) = \left\langle \ln\left(\sum_{d=1}^{D}\left(\frac{x_{id}}{\beta_{jd}}\right)^{\frac{1}{\tau_j}}\right)\right\rangle. \tag{7.75}$$

Due to the convexity feature of the function $\mathcal{G}(\tau_j)$ relative to $\ln \tau_j$, its lower bound can be calculated using a first-order Taylor expansion of $\mathcal{G}(\tau_j)$ for $\ln \tau_j$ at $\ln \tau_{j,0}$ as follows:

$$\mathcal{G}(\tau_j) \geq \mathcal{G}(\tau_{j,0}) + \frac{\partial \mathcal{G}(\tau_j)}{\partial \ln \tau_j}|_{\tau_j=\tau_{j,0}}(\ln \tau_j - \ln \tau_{j,0}), \tag{7.76}$$

$$= \ln\left(\sum_{d=1}^{D}\left(\frac{x_{id}}{\beta_{jd}}\right)^{\frac{1}{\tau_{j,0}}}\right) + \tau_{j,0}\frac{1}{\sum_{d=1}^{D}\left(\frac{x_{id}}{\beta_{jd}}\right)^{\frac{1}{\tau_{j,0}}}}$$

$$\times \frac{-1}{\tau_{j,0}^2}\sum_{d=1}^{D}\left(\frac{x_{id}}{\beta_{jd}}\right)^{\frac{1}{\tau_{j,0}}}\ln\left(\frac{x_{id}}{\beta_{jd}}\right)(\ln \tau_j - \ln \tau_{j,0}),$$

$$= \frac{-\ln \tau_j}{\tau_{j,0}} \frac{\sum_{d=1}^{D} (\frac{x_{id}}{\beta_{jd}})^{\frac{1}{\tau_{j,0}}} \ln (\frac{x_{id}}{\beta_{jd}})}{\sum_{d=1}^{D} (\frac{x_{id}}{\beta_{jd}})^{\frac{1}{\tau_{j,0}}}} + const.$$

References

[1] Foster Provost, Tom Fawcett, Data science and its relationship to big data and data-driven decision making, Big Data 1 (1) (2013) 51–59.

[2] Sharon X. Lee, Geoffrey McLachlan, Saumyadipta Pyne, Application of mixture models to large datasets, in: Big Data Analytics, Springer, 2016, pp. 57–74.

[3] Mohamad Mehdi, Nizar Bouguila, Jamal Bentahar, Trustworthy web service selection using probabilistic models, in: 2012 IEEE 19th International Conference on Web Services, IEEE, 2012, pp. 17–24.

[4] Nizar Bouguila, Djemel Ziou, Using unsupervised learning of a finite Dirichlet mixture model to improve pattern recognition applications, Pattern Recognition Letters 26 (12) (2005) 1916–1925.

[5] B.S. Everitt, An introduction to finite mixture distributions, Statistical Methods in Medical Research 5 (2) (1996) 107–127.

[6] Geoffrey J. McLachlan, David Peel, Finite Mixture Models, John Wiley & Sons, 2004.

[7] Nizar Bouguila, Djemel Ziou, Jean Vaillancourt, Novel mixtures based on the Dirichlet distribution: application to data and image classification, in: International Workshop on Machine Learning and Data Mining in Pattern Recognition, Springer, 2003, pp. 172–181.

[8] Jeffrey D. Banfield, Adrian E. Raftery, Model-based Gaussian and non-Gaussian clustering, Biometrics (1993) 803–821.

[9] Taoufik Bdiri, Nizar Bouguila, Positive vectors clustering using inverted Dirichlet finite mixture models, Expert Systems with Applications 39 (2) (2012) 1869–1882.

[10] Ram B. Jain, Richard Y. Wang, Limitations of maximum likelihood estimation procedures when a majority of the observations are below the limit of detection, Analytical Chemistry 80 (12) (2008) 4767–4772.

[11] Basim Alghabashi, Nizar Bouguila, Finite multi-dimensional generalized gamma mixture model learning based on mml, in: 2018 17th IEEE International Conference on Machine Learning and Applications (ICMLA), IEEE, 2018, pp. 1131–1138.

[12] Dirk Husmeier, The Bayesian evidence scheme for regularizing probability-density estimating neural networks, Neural Computation 12 (11) (2000) 2685–2717.

[13] Taoufik Bdiri, Nizar Bouguila, Djemel Ziou, Variational Bayesian inference for infinite generalized inverted Dirichlet mixtures with feature selection and its application to clustering, Applied Intelligence 44 (3) (2016) 507–525.

[14] Dirk Husmeier, William D. Penny, Stephen J. Roberts, An empirical evaluation of Bayesian sampling with hybrid Monte Carlo for training neural network classifiers, Neural Networks 12 (4–5) (1999) 677–705.

[15] Bjoern Bornkamp, Approximating probability densities by iterated Laplace approximations, Journal of Computational and Graphical Statistics 20 (3) (2011) 656–669.

[16] Lawrence J. Brunner, Albert Y. Lo, et al., Bayes methods for a symmetric unimodal density and its mode, The Annals of Statistics 17 (4) (1989) 1550–1566.

[17] Wentao Fan, Nizar Bouguila, A variational component splitting approach for finite generalized Dirichlet mixture models, in: 2012 International Conference on Communications and Information Technology (ICCIT), IEEE, 2012, pp. 53–57.
[18] Wentao Fan, Nizar Bouguila, Djemel Ziou, Variational learning for finite Dirichlet mixture models and applications, IEEE Transactions on Neural Networks and Learning Systems 23 (5) (2012) 762–774.
[19] Nizar Bouguila, Djemel Ziou, Unsupervised selection of a finite Dirichlet mixture model: an mml-based approach, IEEE Transactions on Knowledge and Data Engineering 18 (8) (2006) 993–1009.
[20] Hagai Attias, Inferring parameters and structure of latent variable models by variational Bayes, in: Proceedings of the Fifteenth Conference on Uncertainty in Artificial Intelligence, 1999, pp. 21–30.
[21] Hagai Attias, A variational Bayesian framework for graphical models, in: Proceedings of the 12th International Conference on Neural Information Processing Systems, 1999, pp. 209–215.
[22] Adrian Corduneanu, Christopher M. Bishop, Variational bayesian model selection for mixture distributions, in: Artificial Intelligence and Statistics, vol. 2001, Morgan Kaufmann, Waltham, MA, 2001, pp. 27–34.
[23] Bo Wang, D.M. Titterington, et al., Convergence properties of a general algorithm for calculating variational Bayesian estimates for a normal mixture model, Bayesian Analysis 1 (3) (2006) 625–650.
[24] Ines Channoufi, Sami Bourouis, Nizar Bouguila, Kamel Hamrouni, Image and video denoising by combining unsupervised bounded generalized Gaussian mixture modeling and spatial information, Multimedia Tools and Applications 77 (19) (2018) 25591–25606.
[25] Nizar Bouguila, Djemel Ziou, Jean Vaillancourt, Unsupervised learning of a finite mixture model based on the Dirichlet distribution and its application, IEEE Transactions on Image Processing 13 (11) (2004) 1533–1543.
[26] Nizar Bouguila, Walid ElGuebaly, A generative model for spatial color image databases categorization, in: 2008 IEEE International Conference on Acoustics, Speech and Signal Processing, IEEE, 2008, pp. 821–824.
[27] Rua Alsuroji, Nizar Bouguila, Nuha Zamzami, Predicting defect-prone software modules using shifted-scaled Dirichlet distribution, in: 2018 First International Conference on Artificial Intelligence for Industries (AI4I), IEEE, 2018, pp. 15–18.
[28] Juansnm José Egozcue, Vera Pawlowsky-Glahn, Simplicial geometry for compositional data, Geological Society, London, Special Publications 264 (1) (2006) 145–159.
[29] Kai Wang Ng, Guo-Liang Tian, Man-Lai Tang, Dirichlet and Related Distributions: Theory, Methods and Applications, vol. 888, John Wiley & Sons, 2011.
[30] Christopher M. Bishop, Pattern Recognition and Machine Learning, Springer, 2006.
[31] Mahieddine M. Ichir, Ali Mohammad-Djafari, A mean field approximation approach to blind source separation with l p priors, in: 2005 13th European Signal Processing Conference, IEEE, 2005, pp. 1–4.
[32] Giorgio Parisi, Statistical Field Theory, Addison–Wesley, 1988.
[33] Wentao Fan, Nizar Bouguila, Online learning of a Dirichlet process mixture of Beta-Liouville distributions via variational inference, IEEE Transactions on Neural Networks and Learning Systems 24 (11) (2013) 1850–1862.
[34] Stephen Boyd, Lieven Vandenberghe, Convex Optimization, Cambridge University Press, 2004.
[35] David J.C. MacKay, Probable networks and plausible predictions—a review of practical Bayesian methods for supervised neural networks, Network Computation in Neural Systems 6 (3) (1995) 469–505.
[36] World Health Organization, et al., World malaria report, 2019.

[37] Malaria datasets | National Library of Medicine, https://lhncbc.nlm.nih.gov/publication/pub9932.
[38] David G. Lowe, Distinctive image features from scale-invariant keypoints, International Journal of Computer Vision 60 (2) (2004) 91–110.
[39] Hieu Nguyen, Muhammad Azam, Nizar Bouguila, Data clustering using variational learning of finite scaled Dirichlet mixture models, in: 2019 IEEE 28th International Symposium on Industrial Electronics (ISIE), IEEE, 2019, pp. 1391–1396.
[40] Koffi Eddy Ihou, Nizar Bouguila, A new latent generalized Dirichlet allocation model for image classification, in: 2017 Seventh International Conference on Image Processing Theory, Tools and Applications (IPTA), IEEE, 2017, pp. 1–6.
[41] Breast cancer statistics, https://www.wcrf.org/dietandcancer/cancer-trends/breast-cancer-statistics, 2019.
[42] Fine needle aspiration (fna) biopsy of the breast, https://www.cancer.org/cancer/breast-cancer/screening-tests-and-early-detection/breast-biopsy/fine-needle-aspiration-biopsy-of-the-breast.html.
[43] Uci machine learning repository, breast dataset, https://archive.ics.uci.edu/ml/datasets/breastcancerwisconsin(original).
[44] Ahmad Taher Azar, Shereen M. El-Metwally, Decision tree classifiers for automated medical diagnosis, Neural Computing and Applications 23 (7–8) (2013) 2387–2403.
[45] Cardiovascular diseases (cvds), http://origin.who.int/cardiovascular_diseases/en/.
[46] Chayakrit Krittanawong, HongJu Zhang, Zhen Wang, Mehmet Aydar, Takeshi Kitai, Artificial intelligence in precision cardiovascular medicine, Journal of the American College of Cardiology 69 (21) (2017) 2657–2664.
[47] Heart disease data set, https://archive.ics.uci.edu/ml/datasets/HeartDisease.
[48] Enrico Blanzieri, Anton Bryl, A survey of learning-based techniques of email spam filtering, Artificial Intelligence Review 29 (1) (2008) 63–92.
[49] Levent Özgür, Tunga Güngör, Optimization of dependency and pruning usage in text classification, Pattern Analysis and Applications 15 (1) (2012) 45–58.
[50] Ola Amayri, Nizar Bouguila, A study of spam filtering using support vector machines, Artificial Intelligence Review 34 (1) (2010) 73–108.
[51] Uci machine learning repository: Spambase data set, https://archive.ics.uci.edu/ml/datasets/spambase.
[52] Yin Zhang, Rong Jin, Zhi-Hua Zhou, Understanding bag-of-words model: a statistical framework, International Journal of Machine Learning and Cybernetics 1 (1–4) (2010) 43–52.
[53] Zhanyu Ma, Arne Leijon, Bayesian estimation of beta mixture models with variational inference, IEEE Transactions on Pattern Analysis and Machine Intelligence 33 (11) (2011) 2160–2173.
[54] Mark William Woolrich, Timothy E. Behrens, Variational Bayes inference of spatial mixture models for segmentation, IEEE Transactions on Medical Imaging 25 (10) (2006) 1380–1391.

CHAPTER 8

From traditional to deep learning: Fault diagnosis for autonomous vehicles

Jing Ren[a], **Mark Green**[b], **and Xishi Huang**[c]
[a]Department of Electrical and Computer Engineering, Ontario Tech University, Oshawa, ON, Canada
[b]Faculty of Science, Ontario Tech University, Oshawa, ON, Canada
[c]RS Opto Tech Ltd., Suzhou, Jiangsu, China

Contents

8.1 Introduction

Autonomous vehicles can be used for many applications where it may be inconvenient, dangerous or impossible to have a human driver present. Equipped with advanced sensors and actuators, autonomous vehicles have the capability to drive, steer, brake and park themselves. Autonomous vehicles normally need more sensors than regular vehicles in order to automatically detect lane boundaries, signs and signals, and static and dynamic obstacles. As technologies improve, autonomous vehicles will be

Learning Control
https://doi.org/10.1016/B978-0-12-822314-7.00013-4

more widespread. It is expected that about half of the automobiles will be autonomous by 2025. Autonomous vehicles have many benefits, the most important of which is road safety. With no human drivers behind the wheel, the number of accidents caused by impaired driving is likely to drop significantly. A 2015 National Highway Traffic Safety Administration report claims 94 percent of traffic accidents happen because of human errors. By eliminating humans from driving operations, autonomous vehicles will make roads safer.

In recent years, the number of sensors and actuators on vehicles has been increasing. As a result, the chance of faults occurring increases. Many fault diagnosis algorithms have been proposed to ensure the safe operation of vehicles. In the literature, the research on fault detection and classification has been active for over 30 years. With more and more advanced control algorithms, there is a growing demand for fault tolerance, which can be achieved not only by improving the reliability of functional units but also by efficient fault detection, isolation and accommodation.

Traditionally, fault diagnosis is reserved for high-end systems such as aircrafts. However, with more advanced techniques and cheaper hardware, fault diagnosis is no longer limited to high-end systems. Consumer products, such as automobiles, are increasingly dependent on microelectronics and mechatronic systems, on-board communication networks, and software, requiring new techniques for achieving fault tolerance [33].

8.2 Traditional fault diagnosis

A fault is defined as an unpermitted deviation of at least one characteristic property or parameter of the system from the acceptable condition [27]. Examples of such malfunctions are the malfunction of an actuator, the breakdown of a sensor or a system component disappearing. Fault diagnosis systems use the idea of redundancy to monitor, locate and identify different faults. The redundancy can be hardware redundancy or control software redundancy. According to [9], fault diagnosis includes three tasks: fault detection, fault isolation, and fault identification. Fault detection is the most basic task of the fault diagnosis, which is used to check whether there is a malfunction or fault in the system and determine the time when the fault occurred. Furthermore, fault isolation determines the location of the faulty component, and fault identification determines the type, nature, and size of the fault. Generally, three major fault diagnosis methods can be found in the literature: model-based methods, signal-based methods and knowledge-based methods [9].

8.2.1 Model-based fault diagnosis

In many fields, software redundancy is more desirable than hardware redundancy due to the extra cost of the hardware. In model-based methods, we need to determine models of the industrial processes or the physical systems. These models can be obtained by using either physical principles or systems identification techniques [20]. Based on the model, we can develop fault diagnosis methods by checking the consistency between the measured outputs of the real systems and the model-based outputs.

The fault can come from sensors, actuators or internal structures. If the measured signals of the sensors are faulty, such as the signals for relative distance, velocity, and acceleration, fatal road accidents might occur. Model-based fault diagnosis has been investigated extensively in the past three decades and some original work has been done recently. In [22], the sliding mode observer was used for probabilistic fault detection by reconstructing the relative acceleration using a longitudinal kinematic model. Based on the predictive diagnostic algorithm, an index that represents the fault ratio quantitatively was proposed for the quantitative evaluation of the fault diagnosis. In [4], a bond graph (BG) tool is used for system modeling, structural analysis and fault diagnosis. This paper provides simple graphical conditions for structural detectability and isolation of plant faults. By combining module theory and BG properties, simple graphical conditions based on the BG model of the system ensure fault detectability and isolation of plant faults. In [6], the authors present two fault tolerant schemes for more reliable spacing control of an autonomous vehicle. The design also includes an estimation of vehicle states and an additive fault with the purpose of achieving highly robust and safe vehicle inter-distance control.

8.2.2 Signal-based fault diagnosis

Signal-based methods make use of measured signals rather than input-output models for fault diagnosis. The faults in the system are embedded in the measured signals. Algorithms are used to extract the features of the signals, and a diagnostic decision is then made based on the signal analysis and prior knowledge of the behavior of the correct systems. Signal-based fault diagnosis methods have a wide application in real-time monitoring and diagnosis for induction motors, power converters, and mechanical components in a system [9].

In recent years, signal-based fault diagnosis continues to evolve with many papers published in this research area. In [1], Abed et al. present a novel approach to the diagnosis of blade faults in an electric thruster motor of unmanned underwater vehicles (UUVs) under stationary operating conditions. The diagnostic approach is based on the use of discrete wavelet transforms (DWTs) as a feature extraction tool and a dynamic neural network for fault classification. The neural network classifies between healthy and faulty conditions of the trolling motor by analyzing the stator current and vibration signals. In [21], the authors investigated the failures of piston and cylinder such as scratching and scuffing which could reduce engine performance and cause the engine breakdowns in their most severe form. They have developed a practical way for the early detection of this fault. To this end, vibration test was conducted on the engine under two states, namely, healthy and faulty.

8.2.3 Knowledge-based fault diagnosis

Model-based methods and signal-based approaches require some prior knowledge about the model or signal patterns. However, in many applications, the model or the signal patterns are unknown. Instead, a large volume of historic data is available. In this case, knowledge-based methods can be utilized. A variety of artificial intelligent techniques can be used to analyze the available historical data of industrial processes, and extract the underlying knowledge which represents the dependence of the system variables. Knowledge-based methods include expert system-based methods, principle component analysis (PCA), partial least squares (PLS), independent component analysis (ICA), statistical pattern classifiers, support vector machine (SVM), artificial neural networks (ANN) and fuzzy logic [8].

In recent years, knowledge-based fault diagnosis methods continue to thrive in different research area. In [25], the authors investigate how to control a quadrotor in such a way that it could determine errors or faults. A fault tolerant control method is developed to use fuzzy logic with battery percentage and degree of ability to hover as the crisp inputs. The fuzzy logic uses five and three membership functions for the battery percentage and degree of ability to hover, respectively. In [5], the nonlinear lateral vehicle model is described by the fuzzy Takagi–Sugeno model. The authors developed a descriptor observer to estimate the system state and faults by ensuring robustness against external disturbances.

8.3 Deep learning for fault diagnosis

Deep learning, which refers to representation learning with multiple layers of nonlinear transformation [3], has been developed to tackle problems in fault diagnosis for different applications. Compared with traditional fault detection methods such as the system identification based method, Deep Neural Network (DNN)-based fault diagnosis can achieve faster and more accurate results [29]. Unlike traditional machine learning methods, DNN consists of many deep layers to extract high-level representations from the original inputs. The output of hidden layers contains the features of different levels. [11] is a joint paper from the major speech recognition laboratories, tackling the first major industrial application of deep learning. Compared with traditional shallow models, which have the problems of lacking expression capacity, using deep learning theory can effectively extract characteristics and accurately recognize the health condition of the components. As a result, fault diagnosis based on deep learning has been an active, productive and promising research field.

DNNs are machine learning techniques, which have the capability to model complex, highly nonlinear relationships between the DNN input and fault classification. Based on prior domain knowledge and a large amount of training data, deep learning can be a powerful and effective tool to detect and classify different complex faults in vehicles, which will significantly improve the quality and speed of the fault decision process. To be specific, deep learning can enable a hierarchical nonlinear learning of high-level features built on top of low-level features to detect which fault(s) are present. Low-level features are the basic details of faults or feature patterns, whereas high-level features are more abstract, that is, high-level features can be obtained by a series of nonlinear transformations through multiple deep layers.

The development of deep learning models for fault diagnosis of autonomous vehicles has been initiated, with more work anticipated in the near future. There are three main DNN types (i.e., Convolutional Neural Networks (CNN), Deep Belief Network (DBN), and Deep Autoencoder (DAE)) that can be used for deep learning and feature extraction. DBNs and DAEs can conduct unsupervised pre-training on the weights, which can ease the difficulty of subsequent supervised training of the deep networks. However, a key problem in DBNs and DAEs is that there are too many weights to train when the inputs are raw signals or their time-frequency representations. In contrast, CNNs can reduce the number of weights to be optimized using the strategies of local receptive field and weight sharing, which can be effective for reducing computational burden during the training process.

8.3.1 Convolutional neural network (CNN)

Convolutional neural network has mainly been used as a classifier for processing images for the last decade. A typical CNN network has an input and an output layer, as well as multiple hidden layers [26]. The hidden layers of a CNN typically consist of a series of convolutional layers. ReLU is the typical activation function, which is normally followed by additional operations such as pooling layers, fully connected layers and normalization layers. Backpropagation is used for error distribution and weight adjustment. A milestone work was done by Yann LeCun et al. in 1989 [16], which used back-propagation to learn the convolution kernel coefficients directly from images of hand-written numbers. This work makes learning completely automatic and the performance is better than manual coefficient adjustment. It is useful for a variety of image recognition problems and image types. This approach became a foundation of modern computer vision. LeNet-5, a pioneering convolutional network by LeCun et al. in 1998 [17], was utilized to recognize hand-written numbers on checks, which were digitized in 32×32 pixel images. The efficient use of convolutional neural networks depends on more layers and larger networks. Therefore, this technique is limited by the computing power and the availability of big data. In Fig. 8.1, we illustrate the structure of CNN.

In recent years, using CNN for fault diagnosis has been gaining strength. In [14], an improved CNN named multi-scale cascade convolutional neural network (MC-CNN) is proposed for the classification information enhancement of input. In MC-CNN, a new layer has been added before the convolutional layers to construct a new signal of more distinguishable information. The effectiveness of MC-CNN is verified by analyzing the application of MC-CNN in bearing fault diagnosis under nonstationary working conditions. In [30], the authors focus on developing a convolutional neural network to learn features directly from the original vibration signals and then diagnose faults. The main faults include imbalance, misalignment of bearing, looseness, and fracture wear of gear components. They show that the one-dimensional convolutional neural network (1-DCNN) model has higher accuracy for fixed-shaft gearbox and planetary gearbox fault diagnosis than that of the traditional diagnostic ones. In [10], the authors propose a hybrid feature model and deep learning-based fault diagnosis

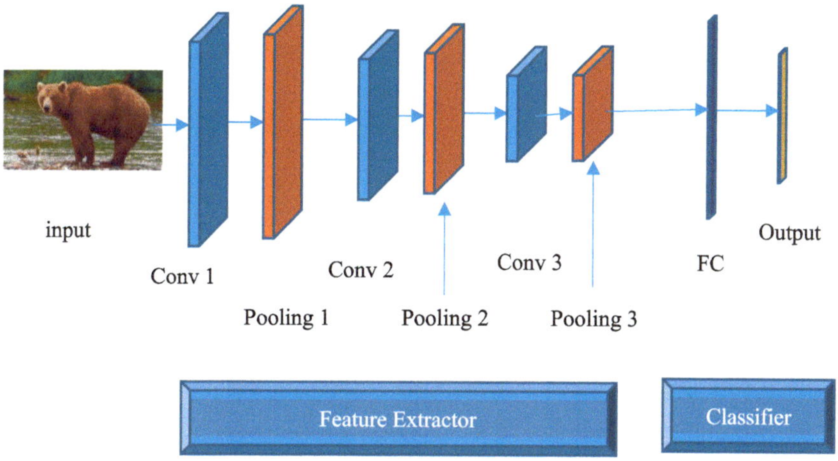

Figure 8.1 The diagram to illustrate CNN.

for Unmanned Aerial Vehicle (UAV) sensors. This work uses Short Time Fourier Transform (STFT) to transform the residual signal to the corresponding time-frequency map. The sensor fault information is extracted by CNN and the fault diagnosis logic between the residuals and the health status is then constructed. Fault diagnosis for planetary gearboxes has been a unique challenge in fault diagnostic field. To date, there are two major approaches to fault diagnosis for planetary gearboxes—vibration analysis and data-driven-based method. The authors in [7] propose an effective and reliable method based on CNN and discrete wavelet transformation (DWT) to identify the fault conditions of planetary gearboxes. The isoelectric line is an important component that connects the steady arm and the drop bracket of catenary in high-speed railway. In [19], the authors propose an automatic fault detection system for the loose strands of the isoelectric line. A convolutional neural network is adopted to extract the isoelectric line features. To accurately and quickly learn these features, an improved feature extraction network, called the isoelectric line network, is presented. Using the images captured from catenary inspection vehicles, the image areas that contain the isoelectric lines are obtained based on the faster region-based convolutional neural network. In [24], the authors proposed a new algorithm for detecting and identifying faults. The most important innovations are image-based processing and classification using deep neural networks. AlexNet is selected as a pre-trained CNN model. In [18], a fault-tolerant control method based on ResNet101 is proposed for the multi-displacement sensor fault of a wheel-legged robot with a new structure. Unlike most methods that only detect a single sensor, the proposed method can detect a large number of sensors simultaneously and rapidly.

8.3.2 Deep autoencoder (DAE)

A deep autoencoder is a neural network that compares its output to its input for learning input data patterns. It has a smaller hidden layer that generates a pattern that can accurately represent the input [28]. A typical autoencoder has two parts: an encoder that maps the input into the code, and a decoder that maps the code to a reconstruction of the original input. Examples of autoencoders include sparse, denoising and contractive autoencoders. Autoencoders are effectively used for solving many applied problems including fault diagnosis. In Fig. 8.2, we illustrate the structure of an autoencoder.

Recently, there have been some papers using DAEs for fault diagnosis for vehicles. In [13], DAEs are developed to automatically and accurately identify bearings faults. The experimental results show that the proposed method can remove the dependence on artificial feature extraction and overcome the limitations of individual deep learning models, which is more effective than other intelligent diagnosis methods. In [31], a new diagnosis technique is proposed to identify large vibration data of multi-level faults in roller bearings. A deep learning network based on stacked autoencoders with hidden layers is exploited for vibration feature representation of roller bearing data, in which the unsupervised learning algorithm is used to reveal the significant properties in the data such as nonlinear, non-stationary properties. Batch learning may waste time and computing resources since

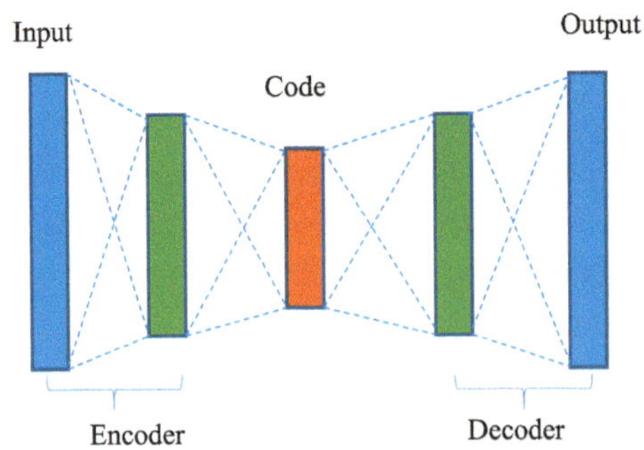

Figure 8.2 The diagram to illustrate autoencoder.

they need to discard the previous learned model and retrain a new model based on the newly acquired data and prior data. In order to overcome this problem, in [15], the authors propose a fault diagnosis method based on class incremental learning without manual feature extraction. Based on a denoising autoencoder, the method obtains the autoencoders using the raw data acquired from healthy states. Only the autoencoder for a new healthy state needs to be trained while the former trained autoencoders are retained.

8.3.3 Deep belief network (DBN)

A DBN is described as a composition of many restricted Boltzmann machines (RBM), which has two layers of feature-detecting units. An RBM is also a special type of Markov random field [32]. The RBM is a generative stochastic artificial neural network that can learn a probability distribution from its input datasets. It can solve the problem of searching the parameter space of deep architectures [2,12]. DBN is one of the hottest topics in the field of neural networks. In recent years, it has shown higher accuracy than some famous existing deep learning methods in image recognition, speech recognition, hand-writing recognition and other classification problems. Different from the conventional shallow learning networks, one of the most important features of DBN is that it facilitates identification of deep patterns, which enables reasoning abilities and makes it possible to capture the deep difference between normal data and faulty data [23,32]. In Fig. 8.3, we illustrate the structure of a restricted Boltzmann machine.

Since the deep belief network was applied to aircraft engine fault diagnosis, more and more scholars have applied deep learning to the field of fault diagnosis and prognosis obtaining many research results. In [32],

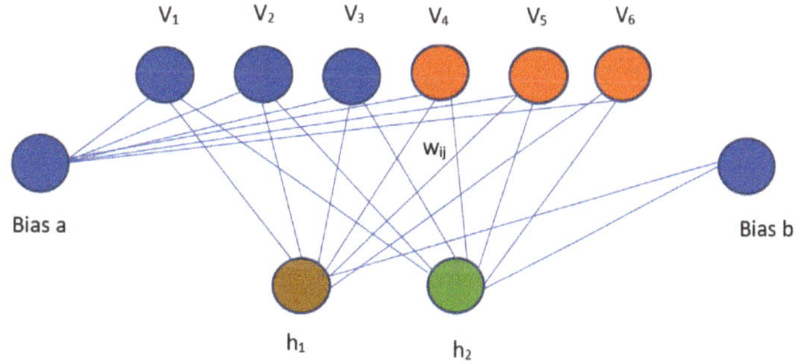

Figure 8.3 The diagram for restricted Boltzmann machine.

the authors proposed a novel DBN-based approach for the fault diagnosis of Vehicle On-board Equipment (VOBEs) in high-speed railways. To capture the complexity and uncertainty of the VOBE faults, they first developed a mathematical model to formulate the fault diagnosis problem in High Speed Railways (HSRs) by the definition of fault evident vectors and reason vectors. Then, they established a deep belief network by the composition of several stacked RBMs. They used the CD-1 (unsupervised) and Greedy-layer algorithm (supervised) to train the developed DBN-BPNN model.

8.4 An example using deep learning for fault detection

In this section, we introduce an example of using deep learning for fault detection.

8.4.1 System dynamics and fault models

8.4.1.1 System dynamics

The target vehicle is a four-wheel independently driven and steered system. The equations of motion for lateral and yaw motion are obtained as follows:

$$M\left(\dot{V}_y+\gamma V_x\right)=2C_x\left(\delta_f-\frac{V_y+l_f r}{V_x}\right)-2C_r\left(\frac{V_y-l_r r}{V_x}\right)+\delta F_y \tag{8.1}$$

$$I\dot{\gamma}=2l_f C_f\left(\delta_f-\frac{V_y+l_f r}{V_x}\right)+2l_r C_r\left(\frac{V_y-l_r r}{V_x}\right)+\delta M_z \tag{8.2}$$

where $\delta_f \in R^{1x1}$ is the front steering angle input from the driver, δF_y and δM_z are control lateral force and control yaw moment, respectively.

From Eqs. (8.1) and (8.2), the state-space equations can be obtained as follows:

$$\dot{x}(t)=Ax(t)+Bu(t) \tag{8.3}$$

$$y(t)=Cx(t)+w(t) \tag{8.4}$$

where $x \in R^{nx1}$ is the state variable, $u \in R^{mx1}$ is the control input, $y \in R^{px1}$ is the measurement output, $w \in R^{px1}$ is the measurement noise.

8.4.1.2 Fault models

In this chapter, we aim to detect three different types of actuator faults. Fault 1 is the combination of an additive fault and a multiplicative fault.

Fault 2 is a multiplicative fault of the motor. Fault 3 is an additive fault of the motor. Therefore, there are four cases: without fault, with multiplicative fault, with additive fault, and with both faults.

The equations for multiplicative and additive faults are as follows:

$$\dot{x}(t) = Ax(t) + B[(I + \alpha)u(t) + f_u] \tag{8.5}$$

where $\alpha \in R^{m \times m}$ represents the multiplicative actuator fault and is a diagonal matrix with the diagonal elements α_{ii}, $i = 1, \ldots, m$, $-1 < \alpha_{ii} < 0$, I is the identity matrix. The additive actuator fault is represented by $f_u \in R^{m \times 1}$. It is assumed that faults are time-invariant.

8.4.2 Deep learning methodology

In this chapter, we first generate simulated data using the vehicle system dynamic model and fault models i.e., the system input signal and the output signals. Figs. 8.4–8.6 show input and output signals. Fig. 8.4 shows the system input signals with faults and without a fault. Fig. 8.5 shows the system output signal lateral velocity $y_1(t)$ for all four cases: without a fault (green), with fault 1 (blue), with fault 2 (yellow), and with fault 3 (pink). Fig. 8.6 shows the system output signal yaw motion $y_2(t)$ for all four cases: without a fault (green), with fault 1 (blue), with fault 2 (yellow), and with fault 3 (pink). In this section, continuous wavelet transform (CWT) is used to transform the 1D input/output time-domain signals $u(t)$, $y_1(t)$, $y_2(t)$

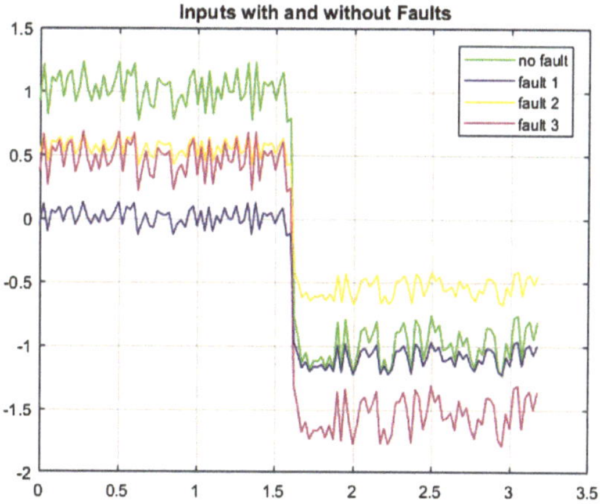

Figure 8.4 The system input signals with faults and without a fault.

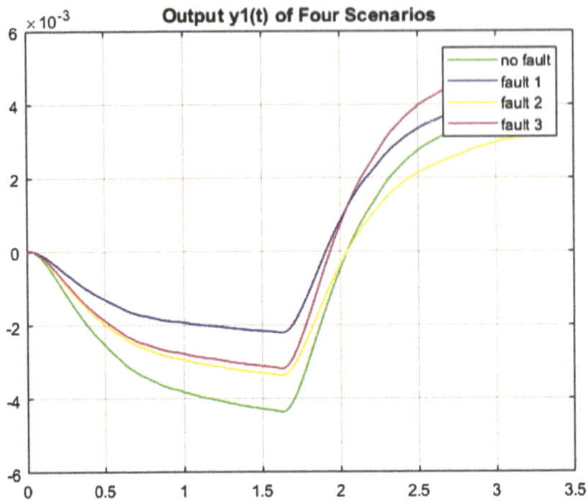

Figure 8.5 The system output signal $y_1(t)$ for all four cases: without a fault (green), with fault 1 (blue), with fault 2 (yellow), and with fault 3 (pink).

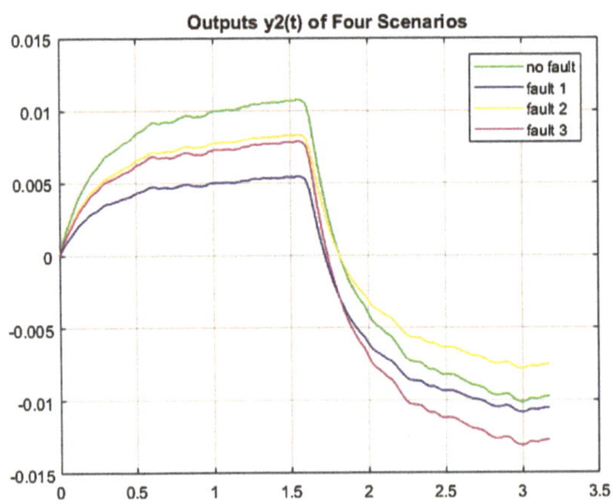

Figure 8.6 The system output signal $y_2(t)$ for all four cases: without a fault (green), with fault 1 (blue), with fault 2 (yellow), and with fault 3 (pink).

into the corresponding 2D time-frequency domain images, respectively, as shown in Fig. 8.7.

Fig. 8.7 shows the schematic of deep convolutional neural network (DCNN) for fault detection and classification. We train the weights of the DCNN which minimize the loss function $J(w)$.

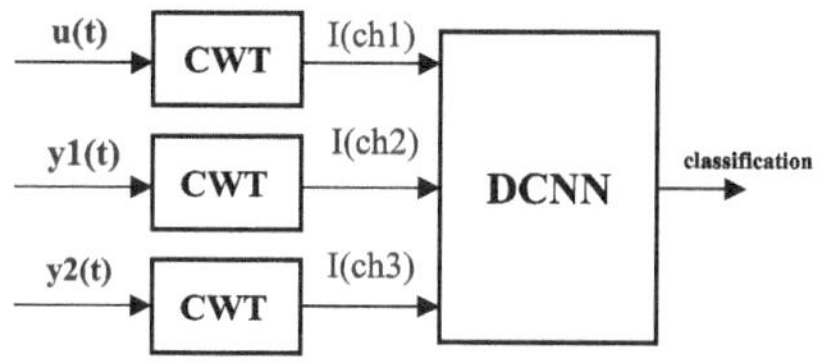

Figure 8.7 The schematic of DCNN fault detection and classification.

Table 8.1 Fault classification results.

Test dataset	Detection correct rate
500 normal signals	99.00%
500 fault 1 signals	100.00%
500 fault 2 signals	100.00%
500 fault 3 signals	98.40%
Average	99.35%

The loss function J(w) is defined as follows:

$$\min_{w} J(w) = \sum_{n=1}^{N} \left(\left| y_n(I_n, w) - y_{tn} \right| \right)^2 \tag{8.6}$$

where n corresponds to the nth sample of training data, N is the number of training samples, w is all the parameters of the deep neural network, y_{tn} is the ground truth label: 0 for normal class, 1 for fault 1, 2 for fault 2, 3 for fault 3.

8.4.3 Fault classification results

In this section, the proposed DCNN fault classification technique is validated. We use a separate test dataset of 2000 images to test the proposed DCNN fault classification approach. The result is shown in Table 8.1. In this study, we have demonstrated that the proposed DCNN technique can effectively detect the faults caused by a mechanical vehicle wheel control failure.

8.5 Conclusion

In this chapter, we have reviewed traditional fault diagnosis methods and three types of new deep learning-based fault diagnosis methods. Although traditional fault diagnosis methods are still effective in many applications, deep learning-based methods have proved to be more and more promising in the field of complex fault diagnosis for autonomous vehicles.

References

[1] W. Abed, et al., Advanced feature extraction and dimensionality reduction for unmanned underwater vehicle fault diagnosis, in: International Conference on Control, Belfast, K, 2016.
[2] Y. Bengio, Learning deep architecture for AI, Foundations and Trends in Machine Learning 2 (1) (2009) 1–55.
[3] Y. Bengio, A. Courville, P. Vincent, Representation learning: a review and new perspectives, IEEE Transactions on Pattern Analysis and Machine Intelligence 35 (8) (Aug. 2013) 1798–1828.
[4] S. Benmoussa, B. Bouamama, R. Merzouki, Bond graph approach for plant fault detection and isolation: application to intelligent autonomous vehicle, IEEE Transactions on Automation Science and Engineering 11 (2) (2014) 585–593.
[5] M.R. Boukhari, et al., Sensor fault tolerant control strategy for autonomous vehicle driving, in: International Multi-Conference on Systems, Signals & Devices, 2016.
[6] M. Boukhari, et al., Two longitudinal fault tolerant control architectures for an autonomous vehicle, Mathematics and Computers in Simulation 156 (2019) 236–253.
[7] R. Chen, et al., Intelligent fault diagnosis method of planetary gearboxes based on convolution neural network and discrete wavelet transform, Computers in Industry (2019).
[8] M.J. Ferreira, C. Santos, J. Monteiro, Cork parquet quality control vision system based on texture segmentation and fuzzy grammar, IEEE Transactions on Industrial Electronics 56 (3) (2009) 756–765.
[9] Z. Gao, C. Cecati, S. Ding, A survey of fault diagnosis and fault-tolerant techniques – Part I: fault diagnosis with model-based and signal-based approaches, IEEE Transactions on Industry Electronics 62 (6) (2015).
[10] D. Guo, et al., A hybrid feature model and deep learning based fault diagnosis for unmanned aerial vehicle sensors, Neurocomputing 319 (2018) 155–163.
[11] G. Hinton, et al., Deep neural networks for acoustic modeling in speech recognition, IEEE Signal Processing Magazine 29 (2012) 82–97.
[12] G. Hinton, A practical guide to training restricted Boltzmann machine, in: Neural Networks: Tricks of the Trade, Springer, Berlin Heidelberg, 2012, pp. 599–619.
[13] K. Horiwaki, Determining manufacturing condition range using a casual quality model and deep learning, in: Annual Conference of the Society of Instrument and Control Engineers of Japan, Hiroshima, Japan, 2019.
[14] W. Huang, et al., An improved deep convolutional neural network with multi-scale information for bearing fault diagnosis, Neurocomputing 359 (2009) 77–92.
[15] J. Kang, et al., A class incremental learning approach based on autoencoder without manual feature extraction for rail vehicle fault diagnosis, in: Prognostics and System Health Management Conference, 2018.
[16] Y. LeCun, et al., Backpropagation Applied to Handwritten Zip Code Recognition, AT&T Bell Laboratories, 1989.
[17] Y. LeCun, et al., Gradient-based learning applied to document recognition, Proceedings of the IEEE 86 (11) (1998) 2278–2324.
[18] Q. Liang, et al., Real-time comprehensive glass container inspection system based on deep learning framework, Electronics Letters 55 (3) (2019) 131–132.
[19] Z. Liu, et al., A high-precision loose strands diagnosis approach for isoelectric line in high-speed railway, IEEE Transactions on Industrial Informatics 14 (3) (2018) 1067–1077.

[20] D. Luo, System Identification and Fault Detection of Complex Systems, PhD Thesis, The University of Central, Florida, 2003.
[21] A. Moosavian, et al., The effect of piston scratching fault on the vibration behavior of an IC engine, Applied Acoustics 126 (2017) 91–100.
[22] K. Oh, et al., Functional perspective-based probabilistic fault detection and diagnostic algorithm for autonomous vehicle using longitudinal kinematic model, Microsystem Technologies 24 (2018) 4527–4537.
[23] A. Oppermann, https://towardsdatascience.com/deep-learning-meets-physics-restricted-boltzmann-machines-part-i-6df5c4918c15.
[24] R. Ozdemir, M. Koc, A quality control application on a smart factory prototype using deep learning methods, in: IEEE International Conference on Computer Sciences and Information Technologies, 2019.
[25] M. Padilla, et al., Fuzzy-based fault tolerant control of micro aerial vehicles (MAV) – a preliminary study, in: IEEE International Conference on Humanoid, Nanotechnology, Information Technology, Communication and Control, Environment and Management, 2017, pp. 91–100.
[26] Prabhu, https://medium.com/@RaghavPrabhu/understanding-of-convolutional-neural-network-cnn-deep-learning-99760835f148.
[27] D. Schrick, Remarks on terminology in the field of supervision, fault detection and diagnosis, in: Proc. IFAC Symp. Fault Detection, Supervision Safety Techn, Process, Hull, U.K., 1997, pp. 959–964.
[28] M. Steward, https://towardsdatascience.com/generating-images-with-autoencoders-77fd3a8dd368.
[29] A. Tellaeche, R. Arana, Three-dimensional machine vision and machine learning algorithms applied to quality control of percussion caps, in: IET Computer Vision, 2010, pp. 117–124.
[30] C. Wu, et al., Intelligent fault diagnosis of rotating machinery based on one-dimensional convolutional neural network, Computers in Industry (2019).
[31] N. Yiethung, et al., Big vibration data diagnosis of bearing fault based on feature representation of autoencoder and optimal LSSVM-CRO classifier model, in: International Conference on System Science and Engineering, 2019, pp. 557–563.
[32] J. Yin, W. Zhao, Fault diagnosis network design for vehicle on-board equipments of highspeed railway: a deep learning approach, Engineering Applications of Artificial Intelligence 56 (2016) 250–259.
[33] Y. Zhang, J. Jiang, Bibliographical review on reconfigurable fault-tolerant control systems, Annual Reviews in Control 32 (2008) 229–252.

CHAPTER 9

Controlling satellites with reaction wheels

Afshin Rahimi
Department of Mechanical, Automotive and Materials Engineering, University of Windsor, Windsor, ON, Canada

Contents

9.1 Introduction

With the ever-increasing demand for smaller satellites, earth observation applications require accurate 3-axis stabilization to enhance the operational envelope and competence of the nanosatellites. The task of controlling the satellite attitude is handled by the attitude control systems (ACS) module on a spacecraft. In general, for a spacecraft that is equipped with reaction wheels (RW), it is assumed that the spacecraft system has a slow-dynamic while the RW has a fast-dynamic. Hence, the actuator dynamics is usually neglected in modeling and analysis. However, high precision for an absolute attitude tracking for a single spacecraft mission necessitates the inclusion of

Learning Control
https://doi.org/10.1016/B978-0-12-822314-7.00014-6

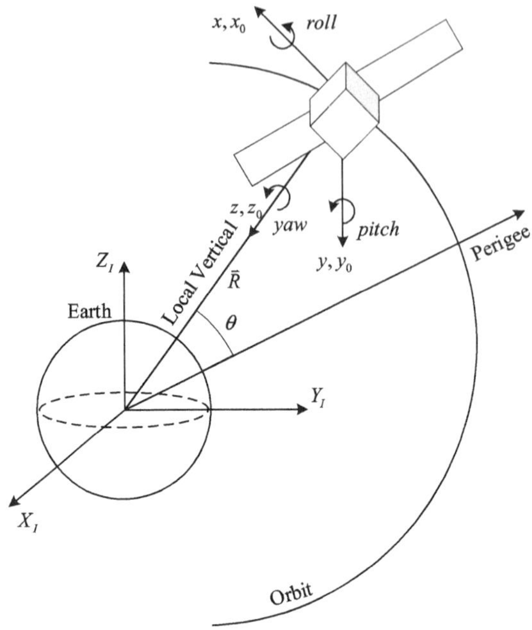

Figure 9.1 The geometry of orbital motion for a rigid spacecraft.

the RW dynamics in the control algorithm formulation. Therefore, in this chapter, a spacecraft dynamics model is developed that integrates the RW dynamics inside. A sliding mode controller is then developed to control the attitude of the spacecraft in 3-axis, and simulations are provided to analyze the performance of the controller for the spacecraft.

9.2 Spacecraft attitude mathematical model

In the first stage of the mathematical model development for a rigid-body spacecraft, the coordinate frames need to be defined.

9.2.1 Coordinate frame

Fig. 9.1 illustrates the three main frames of reference used in this modeling process.

The reference frames of interest include $\Im - X_I Y_I Z_I$, which is an Earth-centered inertial (ECI) reference frame with its origin at the center of the Earth, where Z_I and X_I axes pass through the celestial North Pole and point towards the vernal equinox, respectively; while Y_I completes the right-handed triad. $\mathcal{L} - X_O Y_O Z_O$, is the local vertical local horizontal (LVLH) frame, also referred to as the *orbital frame*. The origin of this frame is at the center of the spacecraft with the X_O and Y_O along the direction of motion, and in the opposite direction of the angular velocity, respectively; while Z_O completes the right-handed triad. $\mathcal{B} - X_B Y_B Z_B$ is the body-fixed coordinate frame, which has its origin at the spacecraft's center of mass. v_{LI}^B denotes a vector of frame $\mathcal{L}$ relative to frame $\Im$, expressed in frame $\mathcal{B}$. $a^{\times} b$ denotes a cross product with $a^{\times}$ as the skew-symmetric of a.

9.2.2 Spacecraft dynamics

A rigid body spacecraft obeys the Euler equation of motion:

$$\dot{H}_{BI}^{I} = \tau_e \tag{9.1}$$

This equation is only valid when expressed in the inertial frame. Now if we express this equation in the body frame, from the *transport theorem* [3]:

$$\dot{H}_{BI}^{B} + \omega_{BI}^{B} \times H_{BI}^{B} = \tau_e \tag{9.2}$$

where $\omega_{BI}^{B}, \tau_e \in \mathbb{R}^{3\times 1}$, and H_{BI}^{B} denote angular velocity, external torques, and total angular momentum of the spacecraft, respectively,

$$H_{BI}^{B} = J\omega_{BI}^{B} + AH_w \tag{9.3}$$

where $A \in \mathbb{R}^{3\times 4}$ is the mapping matrix, J is defined as $J = J_s - AJ_wA^T$ with $J_s \in \mathbb{R}^{3\times 3}$ denoting the moment of inertia (MOI) of the spacecraft, including its actuators. $J_w \in \mathbb{R}^{4\times 4}$ represents the axial MOI of each RW on the diagonal of the matrix. The axial angular momentum of the RWs can be formulated as

$$H_w = J_w(\Omega + A^T\omega_{BI}^{B}) \tag{9.4}$$

where for a RW, $\Omega \in \mathbb{R}^{4\times 1}$ is its axial angular velocity. If we combine Eqs. (9.1) to (9.4), we obtain the attitude dynamics of a fully actuated rigid-body spacecraft controlled by RWs as

$$\begin{aligned}
&\{J\dot{\omega}_{BI}^{B} + A\dot{H}_w\} + \omega_{BI}^{B} \times (J\omega_{BI}^{B} + AH_w) + \tau_e \\
&\longrightarrow J\dot{\omega}_{BI}^{B} = -\omega_{BI}^{B} \times (J\omega_{BI}^{B} + AH_w) - A\dot{H}_w + \tau_e \\
&\longrightarrow J\dot{\omega}_{BI}^{B} = -\omega_{BI}^{B} \times (J_s\omega_{BI}^{B} - AJ_wA^T\omega_{BI}^{B} + AH_w) - A\dot{H}_w + \tau_e \\
&\longrightarrow J\dot{\omega}_{BI}^{B} = -\omega_{BI}^{B} \times (J_s\omega_{BI}^{B} - AJ_wA^T\omega_{BI}^{B} + AJ_w\Omega + AJ_wA^T\omega_{BI}^{B}) - A\dot{H}_w + \tau_e \\
&\longrightarrow J\dot{\omega}_{BI}^{B} = -\omega_{BI}^{B} \times (J_s\omega_{BI}^{B} + AJ_w\Omega) - A\dot{H}_w + \tau_e
\end{aligned}$$

$$J\dot{\omega}_{BI}^{B} = -\omega_{BI}^{B} \times \left(J_s\omega_{BI}^{B} + AJ_w\Omega\right) - A\tau_w + \tau_e \tag{9.5}$$

where the torques that RWs generate are denoted by τ_{RW} given by

$$\tau_w = \dot{H}_w = J_w(\dot{\Omega} + A^T\dot{\omega}_{BI}^{B}) \tag{9.6}$$

In the next section, the attitude kinematics of the spacecraft are discussed.

9.2.3 Attitude kinematics

The kinematic equations for the spacecraft are used to relate the time derivatives of the coordinate frames for the attitude to the angular velocity vector of the spacecraft. There are different representations of these kinematic equations of motion for the spacecraft, including Euler angles, which have gimbal-lock and singularity issues, Cayley–Rodrigues parameters, and a quaternion which is also known in terms of Euler parameters. The unit quaternion is defined by

$$\bar{q} = \begin{bmatrix} q_v \\ q_4 \end{bmatrix} \tag{9.7}$$

where $q_4 \in \mathbb{R}$ and $q_v \in \mathbb{R}^{3\times1} = \left[q_1, q_2, q_3\right]^T$ denote the Euler parameters with $q_v^T q_v + q_4 = 1$. Using these, the nonlinear differential equation for the kinematics of spacecraft can be given by

$$\begin{bmatrix} \dot{q}_v \\ \dot{q}_4 \end{bmatrix} = \frac{1}{2}\begin{bmatrix} q_4 I + q_v^{\times} \\ -q_v^T \end{bmatrix}\omega_{BL}^{B} \tag{9.8}$$

where $I \in \mathbb{R}^{3\times3}$ denotes the identity matrix and $q_v^{\times}$ is given by

$$q_v^{\times} = \begin{bmatrix} 0 & -q_3 & q_2 \\ q_3 & 0 & -q_1 \\ -q_2 & q_1 & 0 \end{bmatrix} \tag{9.9}$$

ω_{BI}^{B} can be formulated as

$$\omega_{BI}^{B} = \omega_{BL}^{B} + \omega_{LI}^{B} \tag{9.10}$$

C_L^B denotes $\mathcal{B}$ to $\mathcal{L}$ transformation matrix given by

$$C_L^B = \left(q_4^2 - q_v^T q_v\right) I + 2q_v q_v^T - 2q_4 q_v^{\times} \tag{9.11}$$

Furthermore, the orbital angular velocity expressed in the body frame is defined as

$$\omega_{LI}^{B} = C_{L}^{B}\begin{bmatrix}0 & -\omega_0 & 0\end{bmatrix}^{T} \tag{9.12}$$

where $\omega_0 = \dot{\theta}$ is the orbital frame's angular velocity which for circular orbits is equal to $\dot{\theta} = \sqrt{\mu_E/R_c^3}$ with μ_E as Earth's gravitational constant and R_p as the spacecraft distance from the center of the Earth given by

$$R_p = R_E + R_S \tag{9.13}$$

Here, R_E and R_S denote the Earth's radius and the altitude of the spacecraft from Earth surface, respectively.

To analyze satellite motion in orbit, it is desired to have $\dot{\omega}_{BL}^{B}$. To obtain this, we have

$$\dot{C}_{L}^{B} = -\left[\omega_{BL}^{B\times}\right]C_{L}^{B} \tag{9.14}$$

Taking a time derivative of (9.10) and applying (9.14) we have

$$\begin{aligned}
\dot{\omega}_{BI}^{B} &= \dot{\omega}_{BL}^{B} + \dot{\omega}_{LI}^{B} \\
&= \dot{\omega}_{BL}^{B} + \frac{d}{dt}\left\{C_{L}^{B}\begin{bmatrix}0 & -\omega_0 & 0\end{bmatrix}^{T}\right\} \\
&= \dot{\omega}_{BL}^{B} + \dot{C}_{L}^{B}\begin{bmatrix}0 & -\omega_0 & 0\end{bmatrix}^{T} \\
&= \dot{\omega}_{BL}^{B} - \left[\omega_{BL}^{B\times}\right]C_{L}^{B}\begin{bmatrix}0 & -\omega_0 & 0\end{bmatrix}^{T} \\
&= \dot{\omega}_{BL}^{B} - \left[\omega_{BL}^{B\times}\right]\omega_{LI}^{B}
\end{aligned} \tag{9.15}$$

However, we are interested in $\dot{\omega}_{BL}^{B}$, hence, by rearranging the terms, we have

$$\dot{\omega}_{BL}^{B} = \dot{\omega}_{BI}^{B} + \left[\omega_{BL}^{B\times}\right]\omega_{LI}^{B} \tag{9.16}$$

9.2.4 External disturbances

In here, gravity gradient (τ_g) and other disturbances constitute the external torques. Hence, $\tau_e = \tau_g + \tau_d$ where τ_g can be modeled as [11]

$$\tau_g = 3\omega_0^2 c_3^{\times} J_s c_3 \tag{9.17}$$

where

$$c_3 = C_{L}^{B}\begin{bmatrix}0 & 0 & 1\end{bmatrix}^{T} \tag{9.18}$$

In addition, τ_d is formulated as [2]

$$\tau_d = \left(\frac{1}{2} + \|\omega_{BL}^{B}\|^2\right)\begin{bmatrix}\sin(0.8t)\\ \cos(0.5t)\\ \cos(0.3t)\end{bmatrix} \tag{9.19}$$

This sounds practical as the extra-terrestrial disturbances are in the form of oscillating waves.

In the next section, attitude tracking is formed to establish a basis for the controller design.

9.3 Attitude tracking

When dealing with tracking a desired rotational motion, we first need to formulate the desired attitude. The attitude of the spacecraft in the desired frame $\mathcal{B}_d$ with respect to $\mathcal{L}$ is denoted by $\left(q_{dv},\ q_{d4}\right) \in \mathbb{R}^3 \times \mathbb{R}$. We denote $\omega_d \in \mathbb{R}^3$ as the angular velocity of $\mathcal{B}_d$ with respect to $\mathcal{L}$, expressed in $\mathcal{B}_d$. Finite constants $c_1,\ c_2 > 0$ exist such that $\|\omega_d\| \leq c_1$ and $\|\dot{\omega}_d\| \leq c_2\ \forall t \geq 0$. The quaternion multiplication of two unit quaternions $q_d = (q_{dv}, q_{d4})$ and $q = (q_v, q_4)$ is denoted by $q_d \odot q$ or $q \otimes q$ and is defined as [12]

$$q_d \otimes q = \begin{bmatrix} q_{d4}q_v + q_4 q_{dv} + \left[q_v^{\times}\right] q_{dv} \\ q_{d4}q_4 - q_{dv}^T q_v \end{bmatrix} \tag{9.20}$$

The quaternion tracking error $q_e = \left(q_{ev}, q_{e4}\right) \in \mathbb{R}^3 \times \mathbb{R}$ is then defined as the relative orientation between $\mathcal{B}$ and $\mathcal{B}_d$. It should be noted that $\overline{q}_e = q_d^{-1} \otimes q$ where q_d is the desired quaternion orientation, q is the current orientation, and $q^{-1} = (-q_v, q_4)$ for $q = (q_v, q_4)$. The quaternion tracking error can then be computed as

$$q_e = q_d \otimes q = \begin{bmatrix} q_{d4}q_v - q_4 q_{dv} + \left[q_v^{\times}\right] q_{dv} \\ q_{d4}q_4 + q_{dv}^T q_v \end{bmatrix} (21) \tag{9.21}$$

which can be separated as

$$\begin{aligned} q_{ve} &= q_{d4}q_v - q_4 q_{dv} + \left[q_v^{\times}\right] q_{dv} \\ q_{4e} &= q_{d4}q_4 + q_{dv}^T q_v \end{aligned} \tag{9.22}$$

The corresponding rotation matrix $C_e = C\left(q_{ve}, q_{4e}\right) \in \mathbb{R}^{3\times3}$ is given by

$$C_e = \left(q_{4e}^2 - q_{ve}^T q_{ve}\right) I + 2 q_{ve} q_{ve}^T - 2 q_{4e}\left[q_{ve}^{\times}\right] \tag{9.23}$$

where $C_e^T C_e = 1$, $\|C_e\| = 1$, $\det(C_e) = 1$, and $\dot{C}_e = -\left[\omega_e^\times\right] C_e$. Next, we define the relative angular velocity $\omega_e \in \mathbb{R}^3$ of $\mathcal{B}$ with respect to $\mathcal{B}_d$ as follows:

$$\omega_e = \omega_{BL}^B - C_e\omega_d \tag{9.24}$$

To derive the error dynamics, we first represent the attitude dynamics in (9.5) in terms of the relative motion of $\mathcal{B}$ in $\mathcal{L}$, which leads to the derivation of Eq. (9.15). From Eqs. (9.5), (9.8), (9.15), (9.22), (9.24), the relative attitude error dynamics and kinematics follow

$$J\dot{\omega}_e = J\left[\left(\omega_{BL}^B \times \omega_{LI}^B\right) + \omega_e^\times C_e\omega_d - C_e\dot{\omega}_d\right] - \omega_{BI}^B \times \left(J_s\omega_{BI}^B + AJ_w\Omega\right) + A\tau_w + \tau_e \tag{9.25}$$

$$\dot{q}_{ve} = \frac{1}{2}\left(q_{4e}I + \left[q_{ve}^\times\right]\right)\omega_e \tag{9.26}$$

$$\dot{q}_{4e} = -\frac{1}{2}q_{ve}^T\omega_e \tag{9.27}$$

Having the attitude tracking, we need to model the actuator dynamics in the next section.

9.4 Actuator dynamics

The model of the RW considered in this section is a simple brushless direct current (BLDC) motor. This model is used for simplicity.

9.4.1 Simple brushless direct current motor

For a RW, torque is produced by changing the speed of rotation for its flywheel that is suspended on ball bearings and driven by a BLDC motor. During the acceleration or deceleration of the flywheel, momentum is produced. The rotation speed of the flywheel (Ω) is proportional to the input voltage to the RW's motor. The torque generated by the motor with respect to armature current $i_a \in \mathbb{R}^{4\times1}$ can be written as [1]

$$\tau_m = K_t i_a \tag{9.28}$$

where $K_t \in \mathbb{R}^{4\times4} = diag(\left[k_1, k_2, k_3, k_4\right])$ denotes the motor torque constant. The induced voltage in the armature for a constant flux (also referred to as the back-electro motive force or BEMF), $e_b \in \mathbb{R}^{4\times1}$, changes directly with Ω. Thus, Faraday's inductance law yields the BEMF as

$$e_b = K_b\Omega \tag{9.29}$$

where $K_b \in \mathbb{R}^{4\times4} = diag([k_{b1}, k_{b2}, k_{b3}, k_{b4}])$ denotes the BEMF constant. The equation of motion for the RW's DC-motor can be obtained via Kirchhoff's voltage law as:

$$L_a\frac{di_a}{dt} + R_a i_a + K_b\Omega = e_b \tag{9.30}$$

where $L_a \in \mathrm{R}^{4\times4} = diag([l_{a1}, l_{a2}, l_{a3}, l_{a4}])$ is the armature inductance (in henry), $R_a \in \mathrm{R}^{4\times4} = diag([r_{a1}, r_{a2}, r_{a3}, r_{a4}])$ and $e_b \in \mathrm{R}^{4\times1}$ denote the armature resistance and the applied armature voltage, respectively. It should be noted that, for simplicity and since the inductance in the RW's armature circuit is negligible, the corresponding term $L_a(di_a/dt)$ is removed from Eq. (9.30) and results in the following linear equation:

$$i_a = R_a^{-1}(e_a - K_b\Omega) \tag{9.31}$$

There are two types of friction in a RW to account for: viscous and Coulomb. The viscous friction is proportional to Ω, while the Coulomb friction is a constant, and its polarity changes with the direction of rotation for the wheel [1]. Hence, the friction for a RW can be modeled as follows:

$$\tau_f = N_c sgn(s) + fs \tag{9.32}$$

where $N_c = 7.06 \times 10^{-4}$ Nm and $f = 1.21 \times 10^{-6}$ Nm/rpm denote the Coulomb and viscous friction coefficients, respectively, and s denotes Ω in revolutions per minute (rpm). Note that Ω is in rad/sec unit. Consequently, the net torque from RW is its motor torque, τ_m, excluding frictional losses τ_f.

$$\tau_{net} = \tau_m - \tau_f \tag{9.33}$$

The RW torque (τ_w) is in the opposite direction of but equal in magnitude to τ_{net}. Hence,

$$\tau_w = -\tau_{net} = \tau_f - \tau_m = N_c sgn(s) + fs - K_t R_a^{-1}(e_a - K_b\Omega) \tag{9.34}$$

The input voltage to the RW, required to get the torque demanded by the spacecraft τ_{w-des}, can be obtained from

$$e_a = K_b\Omega - R_a K_t^{-1}(\tau_{w-des} - \tau_f) \tag{9.35}$$

The flow diagram for the simple BLDC motor model is illustrated in Fig. 9.2. In this figure, it is essential to note the (−1) block which connects

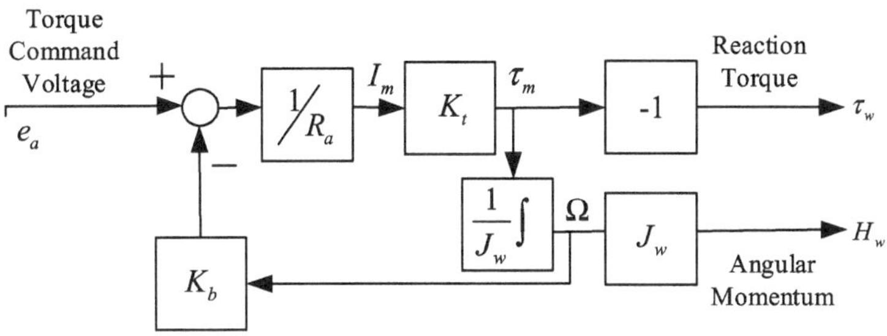

Figure 9.2 Simple brushless direct current motor schematics.

the τ_m signal to τ_w signal. This change in the sign is due to Newton's third law of action/reaction in physics. When applying the control torque, we need to consider this change in sign. If it is considered in the controller, then this block should be removed; otherwise, this block will regulate the sign compensation for the simulations.

There are three different configurations for the RW assembly considered in this study. (1) Three orthogonal wheels [Eq. (9.37)], where there is a dedicated wheel in each principal axis of the assembly. (2) Standard four-wheel configuration [Fig. 9.3(a), Eq. (9.38)] where there is an additional inclined wheel to the three orthogonal configuration that helps control the spacecraft in the case the principal wheels deteriorate, and (3) the pyramid configuration [Fig. 9.3(b), Eq. (9.41)] with an inclined wheel at each corner of the assembly, which offers more redundancy [5].

9.4.2 Mapping matrix

The mapping matrix $A \in \mathbb{R}^{3\times n}$ is used to map each actuator's torque (where the total number of wheels is denoted by n) contribution to the principal axes of the spacecraft body frame as [8,9]

$$\begin{bmatrix} \tau_x & \tau_y & \tau_z \end{bmatrix}^{\mathrm{T}} = A \begin{bmatrix} \tau_{w1} & \tau_{w2} & \tau_{w3} & \tau_{w4} \end{bmatrix}^{\mathrm{T}} \tag{9.36}$$

where $(\cdot)^T$ denotes the transpose of a vector or a matrix. The A matrix will be different for each of the aforementioned configurations as follows.

The mapping matrix A for the three orthogonal wheels configuration is defined as

$$A = \begin{bmatrix} 1 & 0 & 0 & 0 \\ 0 & 1 & 0 & 0 \\ 0 & 0 & 1 & 0 \end{bmatrix} \tag{9.37}$$

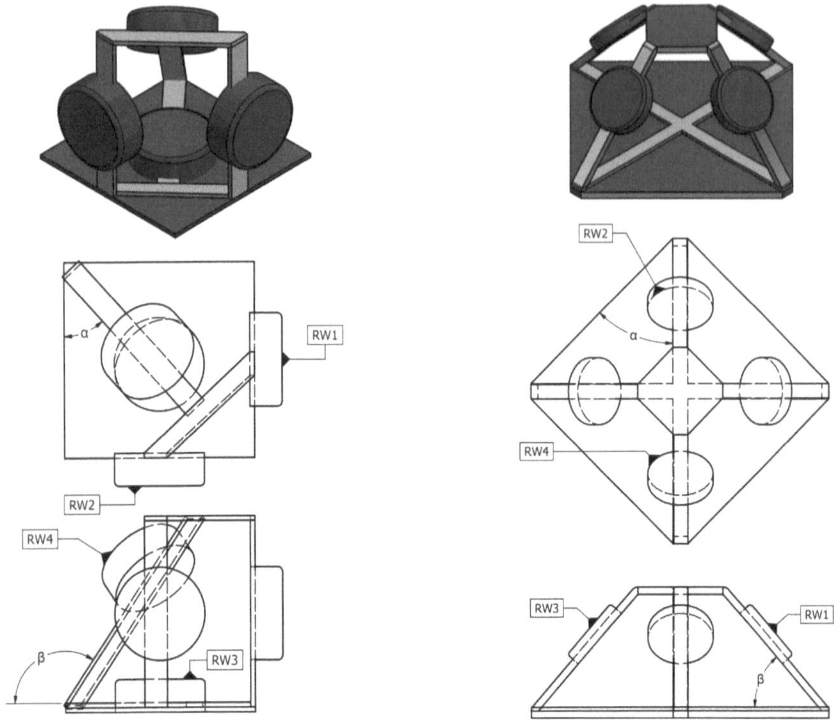

(a) (b)

Figure 9.3 RW assembly; (a) standard four-wheel configuration, (b) pyramid configuration.

This mapping matrix is formed using the fact that there is a wheel in each primary axis, and the fourth column is all zeros because there is no fourth wheel available.

The mapping matrix A for the standard four-wheel configuration is defined as

$$A = \begin{bmatrix} 1 & 0 & 0 & -c\beta s\alpha \\ 0 & 1 & 0 & -c\beta c\alpha \\ 0 & 0 & 1 & s\beta \end{bmatrix} \tag{9.38}$$

where α is the in-plane angle, β is the out-of-plane angle and $s(\cdot)$ and $c(\cdot)$ are $\sin(\cdot)$ and $\cos(\cdot)$, respectively, redundancy [6].

The derivation of this matrix for the first three columns is similar to the orthogonal matrix because there is a wheel in each primary axis. However, for the fourth wheel, we need to decompose its elements in each primary

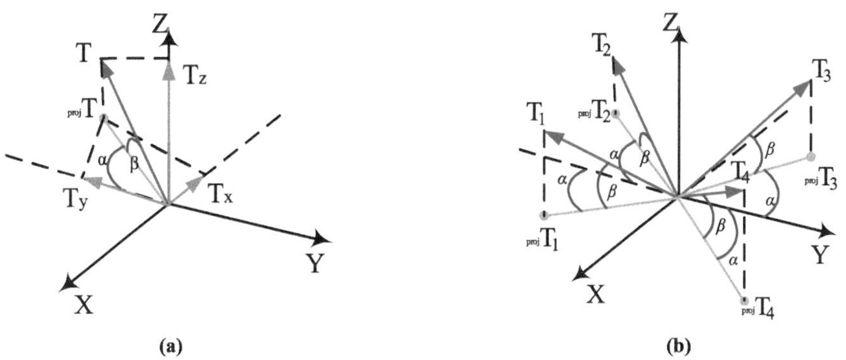

Figure 9.4 Torque decomposition for RW assembly; (a) standard four-wheel, (b) pyramid.

axis. In Fig. 9.4(a), the schematic of the fourth wheel torque with regards to primary axes is shown. The T vector is the torque generated by the oblique reaction wheel. The T_{proj} vector is the projection of T in the XY plane. The angle between these two vectors is shown by β and is called the out-of-plane angle. The other vectors (T_x, T_y, T_z) represent components of T in each of the principal axes.

Having the schematic in Fig. 9.4(a) in order to obtain the fourth column of the matrix in Eq. (9.38), we need to decompose the T vector as follows:

$$\begin{aligned}
&T' = T\cos\beta \\
&T_x = T'\sin\alpha = T\cos\beta\sin\alpha = \textit{considering direction} = -T\cos\beta\sin\alpha \\
&T_y = T'\cos\alpha = T\cos\beta\cos\alpha = \textit{considering direction} = -T\cos\beta\cos\alpha\, T_z \\
&\quad = T\sin\beta
\end{aligned} \tag{9.39}$$

Collecting these terms in a matrix form we get

$$T\begin{bmatrix} -\cos\beta\sin\alpha & -\cos\beta\cos\alpha & \sin\beta \end{bmatrix}^T \tag{9.40}$$

Hence, the fourth column of the matrix is obtained in Eq. (9.38).

The mapping matrix A for the pyramid configuration is defined as

$$A = \begin{bmatrix} c\beta s\alpha & -c\beta s\alpha & -c\beta s\alpha & c\beta s\alpha \\ -c\beta c\alpha & -c\beta c\alpha & c\beta c\alpha & c\beta c\alpha \\ s\beta & s\beta & s\beta & s\beta \end{bmatrix} \tag{9.41}$$

In this configuration, all wheels are contributing to all axes. Hence, all elements of the mapping matrix are non-zero and include trigonometric functions. The schematic of the torque decomposition for pyramid configuration is shown in Fig. 9.4(b). The same convention is used for α and β angles here as was in Fig. 9.4(a). The only difference is that instead of only one oblique wheel, there are four oblique wheels in this arrangement. α is the angle between the projection of each torque to the XY plan and the Y axis for all wheels while β is the angle between actual torque and its projection in the XY plane. The decomposition for wheel 1 can be obtained as follows:

$$\begin{aligned}
T_1' &= T_1 \cos\beta \\
T_{1x} &= T_1' \sin\alpha = T_1 \cos\beta \sin\alpha = \textit{considering direction} = T_1 \cos\beta \sin\alpha \\
T_{1y} &= T_1' \cos\alpha = T_1 \cos\beta \cos\alpha = \textit{considering direction} = -T \cos\beta \cos\alpha \, T_{1z} \\
&= T \sin\beta
\end{aligned} \tag{9.42}$$

Hence, the first column of the mapping matrix reads

$$T_1 \begin{bmatrix} \cos\beta \sin\alpha & -\cos\beta \cos\alpha & \sin\beta \end{bmatrix}^T \tag{9.43}$$

For the rest of the wheels, the same procedure can be followed to arrive at Eq. (9.41).

9.4.3 Reaction wheel parameters

The following BEMF/torque constants are obtained from [4] (Table 9.1). It should be noted that in the simulations in this section, it is assumed that RWs work within the dead-zone/saturation ranges. Hence, no saturation or under-actuation occurs during maneuvers.

Table 9.1 Reaction wheel parameters.

Wheel	BEMF constant [V-sec/rad]	Torque constant [Nm/A]	R_a [Ohm]
1	0.0082	0.0082	0.6
2	0.0080	0.0080	0.6
3	0.0071	0.0071	0.6
4	0.0075	0.0075	0.6

9.5 Attitude control law

This section provides an overview for the development of the attitude control law employed in this study.

9.5.1 Basics of variable structure control

Variable structure control (VSC) is a subset of control theory in which the 'control law' is changed during the simulation. This change is reliant on predefined rules that deal with changes in the system states. As an example, consider the following linear time-invariant (LTI) system in state-space form:

$$\begin{bmatrix} \dot{X}_1 \\ \dot{X}_2 \end{bmatrix} = \begin{bmatrix} 0 & 1 \\ 0 & 0 \end{bmatrix} \begin{bmatrix} X_1 \\ X_2 \end{bmatrix} + \begin{bmatrix} 0 \\ 1 \end{bmatrix} U \tag{9.44}$$

where $X \in \mathbb{R}^2 = [X_1, \ X_2]^T$ is the states vector, and U is a scalar control input. The next step is to define the sliding surface/manifold. We can define the sliding surface as a linear function

$$S = pX_1 + X_2 \tag{9.45}$$

where p is a positive design scalar that impacts the dynamic performance of the sliding surface. Once the sliding manifold is formed, one needs to define the control law

$$U = -\eta sgn(S) = \begin{cases} -1 & \text{if } S \geq 0 \\ 1 & \text{if } S < 0 \end{cases} \tag{9.46}$$

where η is a positive design scalar, and the speed at which the sliding surface can be reached is directly proportional to this parameter. Fig. 9.5 illustrates the phase portrait of the closed-loop system in (9.44) controlled by (9.46) under (9.45). In this simulation $p = 1$ and $\eta = 2$. Different initial conditions are considered to evaluate the performance of the control law. In Fig. 9.5, there are four different initial conditions illustrated with different markers. The ultimate goal for the system with any of the initial conditions is to move to the origin of the phase portrait plane $(0, 0)$, which the figure shows is happening for any of the system's initial conditions.

The inclined line represents $S = 0$ and is known as the sliding manifold. There are four regions in the phase plane divided by S. These four regions

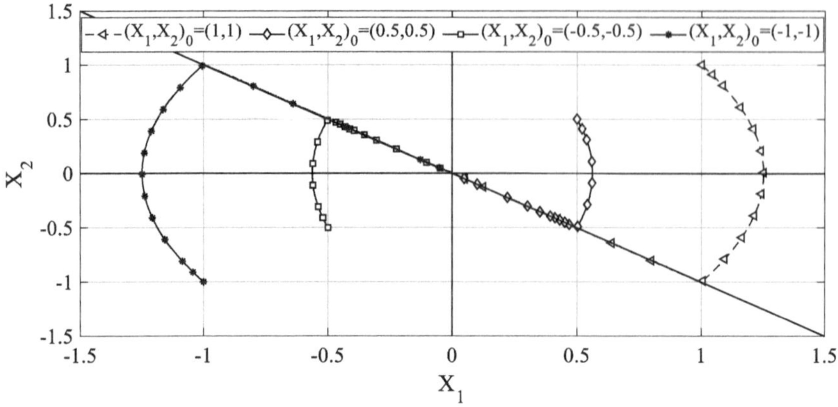

Figure 9.5 Phase portrait of the double integrator under VSC.

are

$$\begin{array}{lll} \text{I: } X_1 > 0, S > 0 & \text{and} & \text{II: } X_1 > 0, S < 0 \\ \text{III: } X_1 < 0, S < 0 & \text{and} & \text{IV: } X_1 < 0, S > 0 \end{array} \tag{9.47}$$

The control law works in such a way that for a given initial condition U drives the system's trajectory towards $S = 0$ line. For the X_2 values that satisfy $p\,|\eta| < K_2$. By choosing a candidate Lyapunov function as $V = S^2/2$ we need to prove $\dot{V} < 0$ for stability.

$$\begin{aligned} \dot{V} &= S\dot{S} \\ &= S\left[K_1\dot{X}_1 + \dot{X}_2\right] \\ &= S\left[K_1X_2 - K_2 sgn\,(S)\right] \\ &= \left[S \times sgn\,(S)\right]\left[\frac{\mathrm{p}X_2}{sgn\,(S)} - K_2\right] \\ &= |S|\,(p\,|X_2| - \eta) \end{aligned} \tag{9.48}$$

And

$$\lim_{S \to 0^+} \dot{S} < 0 \quad \text{and} \quad \lim_{S \to 0^-} \dot{S} > 0 \tag{9.49}$$

Therefore, when $K_1\,|X_2| < K_2$ the system trajectories point towards the $S = 0$ line regardless of which side of the line they are at. The $S\dot{S} < 0$ condition is referred to as "reachability condition".

The trajectory of the system that consists of $S = 0$, can be rearranged to

$$S = pX_1 + X_2$$

$$\begin{aligned} &= pX_1 + \dot{X}_1 \\ &\rightarrow \dot{X}_1 = -pX_1 \end{aligned} \tag{9.50}$$

This equation is a first-order decay trajectory that results in states sliding along the sliding surface ($S = 0$) and eventually converging to the origin. Such dynamic behavior is also referred to as "ideal sliding mode". It should be noted that during the sliding motion, the behavior of the uncontrolled system is dominated by the lower-order dynamics. Hence, the only thing that the control scheme does is to ensure that the sliding surface is reached on the system's trajectory, and the Lyapunov condition is satisfied. From there, the system dynamics will converge to the origin.

The finite-time (t_r) in which the system trajectory converges to the sliding surface (reaching time) can be obtained by

$$\begin{aligned} &S\dot{S} \leq -\eta\,|S| \\ &\rightarrow \int_0^{t_r} \frac{S}{|S|}\dot{S}d\tau \leq \int_0^{t_r} -\eta d\tau \\ &\rightarrow |S(t_r)| - |S(0)| \leq -\eta t_r \end{aligned} \tag{9.51}$$

Since $|S(t_r)| = 0$, the reaching time is equal to

$$t_r \leq \frac{|S(0)|}{K_2} \tag{9.52}$$

9.5.2 Design of sliding manifold

In this section, the control law used to stabilize the attitude of the rigid body spacecraft with an attitude tracking capability is presented based on the work in [4]. To develop the control scheme, we first need to define the following sliding manifold:

$$s = \omega_B + \lambda q_v \tag{9.53}$$

From the dynamics of the spacecraft we have

$$\begin{aligned} \left[J_s - AJ_wA^T\right]\dot{\omega}_B &= -\omega_B^{\times}\left[J_s\omega_B + AJ_w\omega_w\right] + A\tau_w + \tau_d \\ \left[J_s - AJ_wA^T\right]\dot{s} &= A\tau_w + \Delta - \lambda q_v \\ \Delta = -\omega_B^{\times}\left[J_s\omega_B + AJ_w\omega_w\right] &+ \tau_d + \lambda\left[J_s - AJ_wA^T\right]\dot{q}_v + \lambda q_v \end{aligned} \tag{9.54}$$

The system nonlinearities and external disturbances are lumped into a single entity Δ. This uncertainty is not considered in the control algorithm, but it is assumed to be bounded within

$$\|\Delta\| \leq p_1 \|\omega\| + p_2 \|q\| + p_3 \leq \eta \tag{9.55}$$

9.5.3 Control law

The feedback control law is chosen as [4]

$$\begin{aligned} \tau_{w-des} &= -A^T \left[k_1 + (\rho + \eta) \frac{1}{\|s\| + \delta} \right] s \\ \rho &= \sigma \left[p_1 \|\omega\| + p_2 \|q\| + p_3 \right] \end{aligned} \tag{9.56}$$

where k_1, p_1, p_2, p_3, η and δ are positive scalars. The value of σ is determined using the following adaptive law:

$$\dot{\sigma} = b_1 \left[p_1 \|\omega\| + p_2 \|q\| + p_3 \right] \frac{\|s\|^2}{\|s\| + \delta} - b_2 \sigma \tag{9.57}$$

Using the desired torque, and considering negligible friction loss τ_m, from Eq. (9.35) the voltage required is computed and commanded to the wheel by the following relation:

$$e_a = R_a K_t^{-1} J_w \tau_{w-des} + K_b \Omega \tag{9.58}$$

9.5.4 Stability analysis

If we choose the Lyapunov function as follows [4]:

$$V = \frac{1}{2} s^T J s + \frac{1}{2b_1} \sigma^2 + \lambda \left[q_v^T q_v + (1 - q_4^2) \right] \tag{9.59}$$

Taking the first derivative of V along the trajectories of the system and substituting Eq. (9.54) yields

$$\dot{V} = s^T \left[A\tau_w + \Delta \right] + \frac{1}{b_1} \sigma \dot{\sigma} - \lambda^2 q_v^T q_v \tag{9.60}$$

Substituting the control law in Eq. (9.56) and the adaptive law in Eq. (9.57) in Eq. (9.60)

$$\begin{aligned} \dot{V} = s^T &\left[-k_1 s - (\rho + \eta) \frac{s}{s + \delta} + \Delta \right] \\ &+ \frac{1}{b_1} \sigma \left[b_1 \left[p_1 \|\omega\| + p_2 \|q\| + p_3 \right] \frac{\|s\|^2}{\|s\| + \delta} - b_2 \sigma \right] - \lambda^2 q_v^T q_v \end{aligned}$$

$$
\begin{aligned}
&\leq -k_1 \|s\|^2 - (\rho + \eta)\frac{\|s\|^2}{\|s\| + \delta} + [p_1 \|\omega\| + p_2 \|q\| + p_3]\left\{\|s\| + \sigma\frac{\|s\|^2}{\|s\| + \delta}\right\} \\
&\quad - \frac{b_2}{b_1}\sigma^2 - \lambda^2 q_v^T q_v \\
&\leq -k_1 \|s\|^2 - \eta\frac{\|s\|^2}{\|s\| + \delta} + [p_1 \|\omega\| + p_2 \|q\| + p_3] \|s\| - \frac{b_2}{b_1}\sigma^2 - \lambda^2 q_v^T q_v
\end{aligned} \tag{9.61}
$$

Using the fact that $\frac{\|s\|}{\|s\|+\delta} \leq 1$ and Eq. (9.55), we get

$$
\dot{V} \leq -k_1 \|s\|^2 - \eta \|s\| - \frac{b_2}{b_1}\sigma^2 - \lambda^2 q_v^T q_v < 0 \tag{9.62}
$$

Thus, using Barbalat's lemma, we can prove that all the states converge to zero as the time approaches infinity.

9.6 Performance analysis

In this section, simulation results for attitude control of the rigid body spacecraft dynamics developed in the earlier sections are presented. In these simulations, the objective is to lead the spacecraft from an initial attitude to the desired attitude and keep the system stable at that desired orientation. The spacecraft model parameters are given in Table 9.2 [7].

In terms of control gains, Table 9.3 provides the gains used for controlling the attitude of the spacecraft.

Finally, for the simulations, we choose initial conditions and desired orientation as presented in Table 9.4. These values are given in angles;

Table 9.2 Simulation model parameters.

Component	Parameter	Value
Orbit	R_E	6378 [km]
	R_s	500 [km]
	$R_p = R_E + R_s$	6878 [km]
	μ_E	398600 [km^3 s^{-2}]
Satellite	Dimensions	$0.1 \times 0.1 \times 0.1$ [m^3]
	J_{xx}	0.015 [kg m^2]
	J_{yy}	0.017 [kg m^2]
	J_{zz}	0.020 [kg m^2]
Reaction wheels	J_w	1×10^{-5} [kg m^2]

Table 9.3 Control gains for satellite attitude control.

Parameter	Value	Parameter	Value
p_1	1	b_2	0.001
p_2	1	k_1	0.06
p_3	1	λ	1
b_1	300	δ	0.001

Table 9.4 Initial and final conditions for simulations.

Condition	Parameter	Value
Initial	Roll (ϕ_0)	$-90°$
	Pitch (θ_0)	45°
	Yaw (ψ_0)	90°
	$[\omega_{01}, \omega_{02}, \omega_{03}]$	[0, 0, 0] rad/sec
Desired	Roll (ϕ_d)	0°
	Pitch (θ_d)	0°
	Yaw (ψ_d)	0°
	$[\omega_{d1}, \omega_{d2}, \omega_{d3}]$	[0, 0, 0] rad/sec

however, kinematics and dynamics equations work with quaternions. In order to resolve this matter, a function is used to convert Euler angles to quaternion representations, and then those quaternions are fed into the system [10].

It is important to note that the angles given in Table 9.4 are in the orbital frame $\mathcal{L}$. The Simulink model of this simulation is illustrated in Fig. 9.6. Note: both α and β angles for the pyramid configuration mapping matrix are set to 45° in the simulations presented in this section. Simulations are executed on an inter Core with an i7-8650U processor that houses 1.90 GHz and 2.11 GHz of computing power. In addition, the system is equipped with 16 GB of RAM with solid-state drive (SSD) technology storage. The total duration of the simulations is 20 seconds, and the solver for these simulations is the Dormand–Prince variable step ordinary differential equation solver (ODE45).

Simulations are done for the same scenario as described in Table 9.4 but for all three reaction wheel configurations. In Fig. 9.7 for orthogonal case results are presented, it is important to note that one of the wheels is inactive because only three wheels are actively controlling the system. In other words, because the mapping matrix, in this case, has one of its columns as all zeros. Fig. 9.8 and Fig. 9.9 present the simulation results for

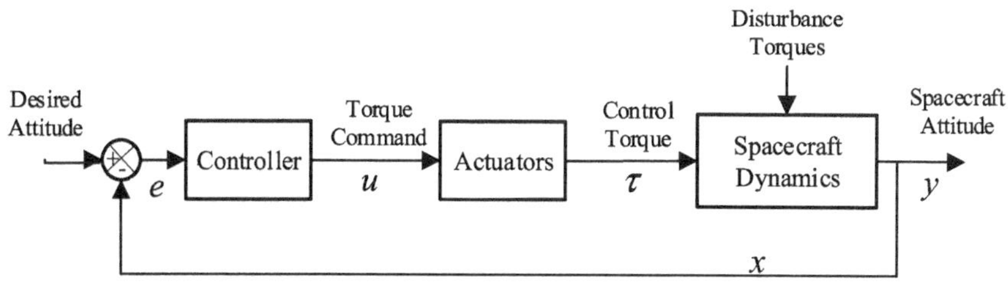

Figure 9.6 Simulation setup in the Simulink for the satellite attitude control.

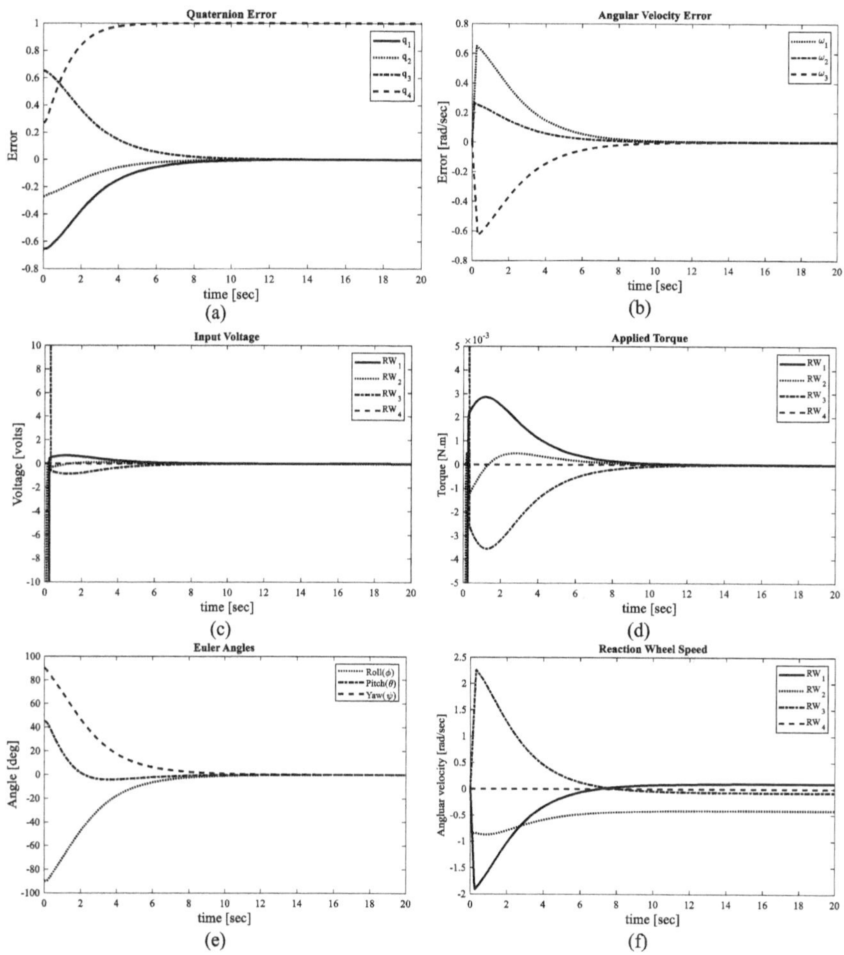

Figure 9.7 Simulation results for orthogonal wheel configuration: (a) quaternion error, (b) angular velocity error, (c) input voltage, (d) applied torque, (e) Euler angles, (f) reaction wheel angular speed.

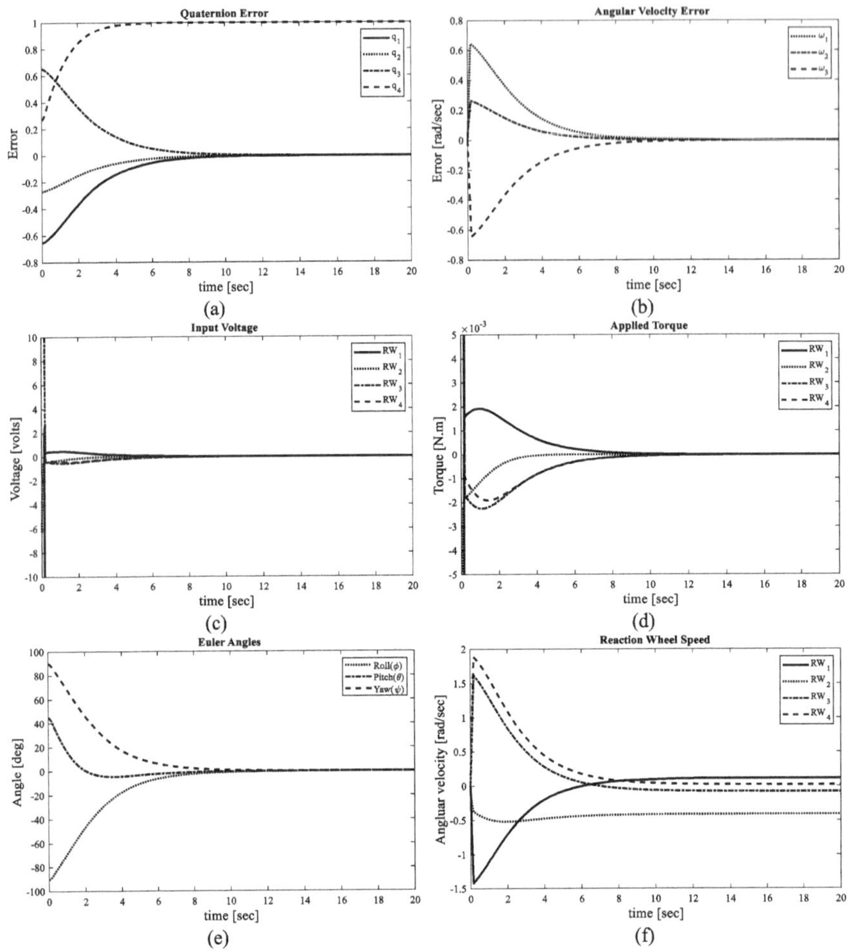

Figure 9.8 Simulation results for standard four-wheel configuration: (a) quaternion error, (b) angular velocity error, (c) input voltage, (d) applied torque, (e) Euler angles, (f) reaction wheel angular speed.

the standard four-wheel and the pyramid configurations, respectively. It can be seen that all four wheels are actively contributing to the control of the system. For all cases, in sub-figure (a), the quaternion vector goes to zero, which means the error converges to zero, and the desired orientation is achieved. In sub-figure (b), angular velocity converges to zero, which means the orientation of the satellite has converged to the desired orientation and is stable. The reaction wheel input voltage is illustrated in sub-figure (c) followed by the torque generated from that voltage in sub-figure (d). It

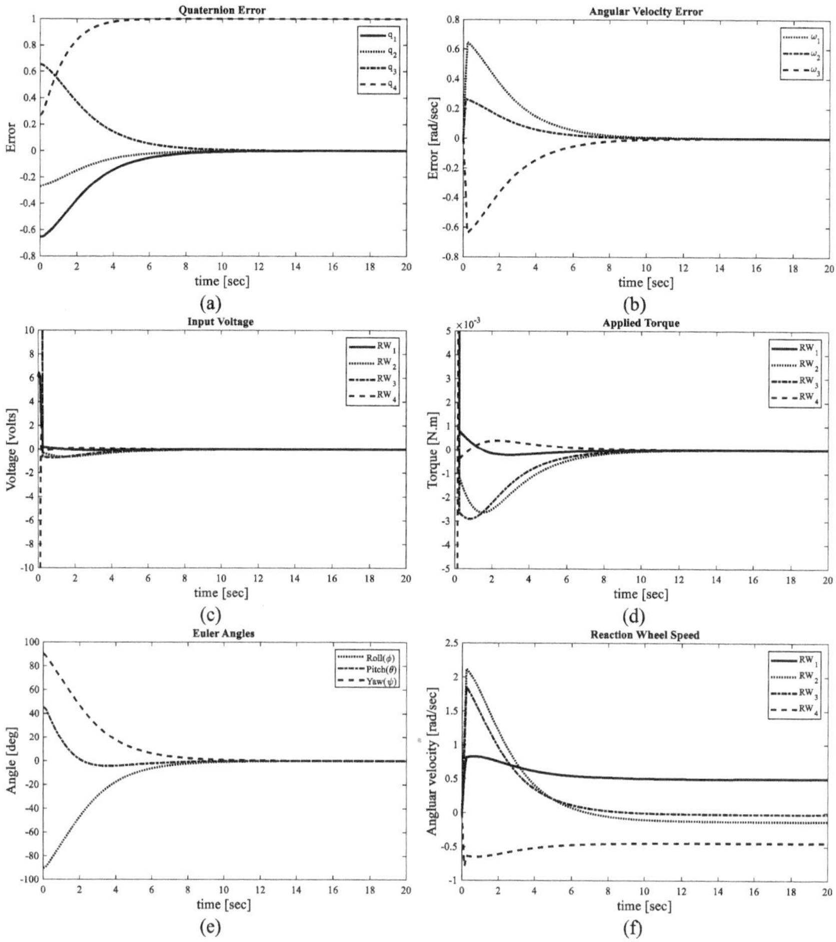

Figure 9.9 Simulation results for pyramid configuration: (a) quaternion error, (b) angular velocity error, (c) input voltage, (d) applied torque, (e) Euler angles, (f) reaction wheel angular speed.

can be seen that, as the orientation of the satellite converges to the desired orientation, the input voltage to the reaction wheels and consequently the generated torque by these actuators converges to zero. However, due to external disturbances applied to the satellite from gravity gradient torque, there will always be a minimal component of these forces acting on the system requiring compensation from the actuators for that disturbance to keep the satellite in its desired orientation. Euler angles corresponding to the satellite orientation are shown in sub-figure (e), where it is more feasible to see the trend in which the system goes from initial conditions to the desired orientation. Finally, in sub-figure (f), reaction wheels angular speeds are presented. This sub-figure has a similar trend as the one for the input voltage to the reaction wheel assembly. This similarity is caused by the fact that reaction wheels angular speeds has a direct relationship with reaction wheels input voltage and applied torque [Eqs. (9.34), (9.35)].

9.7 Conclusions

In this study, a detailed approach was provided on how to develop the analytical model of a rigid-body spacecraft, integrate reaction wheel dynamics to it, develop a sliding mode controller for its attitude and analyze the performance of the system for various reaction wheel assembly configurations.

References

[1] B. Bialke, High fidelity mathematical modeling of reaction wheel performance, in: 1998 Annual AAS Rocky Mountain Guidance and Control Conference, Advances in the Astronautical Sciences, 1998, pp. 483–496.
[2] W. Cai, X. Liao, D.Y. Song, Indirect robust adaptive fault-tolerant control for attitude tracking of spacecraft, Journal of Guidance, Control, and Dynamics 31 (5) (2008) 1456–1463, https://doi.org/10.2514/1.31158.
[3] A.H.J. De Ruiter, C.J. Damaren, J.R. Forbes, Spacecraft Dynamics and Control An Introduction, 1st ed., John Wiley & Sons Ltd., 2013.
[4] Godard, K.D. Kumar, A.-M. Zou, A novel single thruster control strategy for spacecraft attitude stabilization, Acta Astronautica 86 (2013) 55–67, https://doi.org/10.1016/j.actaastro.2012.12.018.
[5] A. Rahimi, K.D. Kumar, H. Alighanbari, Particle swarm optimization aided Kalman filter for fault detection, isolation and identification of a reaction wheel unit, CASI ASTRO 12 (2012).
[6] A. Rahimi, K.D. Kumar, H. Alighanbari, Enhanced adaptive unscented Kalman filter for reaction wheels, IEEE Transactions on Aerospace and Electronic Systems 51 (2) (2015) 1568–1575, https://doi.org/10.1109/TAES.2014.130766.
[7] A. Rahimi, K.D. Kumar, H. Alighanbari, Fault estimation of satellite reaction wheels using covariance based adaptive unscented Kalman filter, Acta Astronautica 134 (2017) 159–169, https://doi.org/10.1016/j.actaastro.2017.02.003.
[8] A. Rahimi, K.D. Kumar, H. Alighanbari, Fault isolation of reaction wheels for satellite attitude control, IEEE Transactions on Aerospace and Electronic Systems 56 (1) (2020) 610–629, https://doi.org/10.1109/TAES.2019.2946665.
[9] A. Rahimi, K.D. Kumar, H. Alighanbari, Failure prognosis for satellite reaction wheels using Kalman filter and particle filter, Journal of Guidance, Control, and Dynamics 43 (3) (2020) 585–588, https://doi.org/10.2514/1.G004616.
[10] A. Rahimi, A. Saadat, Fault isolation of reaction wheels onboard 3-axis controlled in-orbit satellite using ensemble machine learning techniques, in: The International Conference on Aerospace System Science and Engineering, 2019.
[11] H. Schaub, J.L. Junkins, Analytical Mechanics of Space Systems, AIAA Education Series, 2003, 11,34,37,112,113.
[12] M.D. Shuster, A survey of attitude representations, The Journal of the Astronautical Sciences 41 (4) (1993) 439–517, (484).

CHAPTER 10

Vision dynamics-based learning control

Sorin Grigorescu[a,b]
[a]Robotics, Vision and Control (ROVIS), Transilvania University of Brasov, Brasov, Romania
[b]Artificial Intelligence, Elektrobit Automotive, Brasov, Romania

Contents

10.1 Introduction

In the last couple of years, autonomous vehicles began to migrate from laboratory development and testing conditions to driving on public roads. An autonomous vehicle is an intelligent agent which observes its environment, makes decisions and performs actions based on these decisions. The driving functions map sensory input to control output and are implemented either as modular perception-planning-action pipelines [1], End2End [2,3] or Deep Reinforcement Learning (DRL) [4] systems. Visual perception methods, mainly designed in the computational intelligence community, are often decoupled from the low-level control algorithms developed in the automatic control community. Although efforts to bring perception and control closer together have been made in the area of visual servoing, usually such systems are decoupled and do not take into account the intrinsic dependencies between the two components.

In this work, we contribute at bridging the gap between visual perception and control by introducing the concept of *Vision Dynamics* as a learning control paradigm for autonomous vehicles. The term "dynamic vision" was first coined by Dickmanns in his work on visual road and ego-state estimation for autonomous driving [5]. Our method is based on the synergy between a constrained NMPC and a vision dynamics model implemented within the layers of a Deep Neural Network (DNN). The network

Learning Control
https://doi.org/10.1016/B978-0-12-822314-7.00015-8

is trained in an imitation learning setup, with a modified version of the Q-learning algorithm [6].

The combination of model-based control with data driven techniques has been considered previously, for example in [7], using model predictive path integral control (MPPI) for the task of aggressive driving, as well as in [3], using imitation learning for computing the steering and throttle commands in an End2End fashion. While the authors demonstrate impressive results, both approaches take into consideration an obstacle free race track, where predefined boundaries are used to compute the optimal trajectory of the vehicle. Due to these requirements, their robotic system can only operate in a partially controlled environment, with no static or dynamic obstacles, which limits the applicability of their approach to self-driving cars.

Traditional controllers make use of an *a priori* model composed of fixed parameters. When robots or other autonomous systems are used in complex environments, such as driving, traditional controllers cannot foresee every possible situation that the system has to cope with. Unlike controllers with fixed parameters, learning controllers make use of training information to learn their models over time. With every gathered batch of training data, the approximation of the true system model becomes more accurate, thus enabling both model flexibility and consistent uncertainty estimates [8].

In previous work, learning controllers have been introduced based on simple function approximators, such as Gaussian Process (GP) modeling [9–12], or support vector regression [13]. Learning based unconstrained and constrained NMPC controller have been proposed in [11] and [12], respectively. Given a trajectory path, both approaches use a simple *a priori* model and a GP learned disturbance model for the path tracking control of a mobile robot. The experimental environment comprises of fixed obstacles, without any moving objects. Unlike these examples, in our work we primarily model the dynamics of the scene based on visual information processed by different deep neural network architectures. Neural networks have been previously used to model plant dynamics [14,15]. In the last years, they have been also applied to model-based control of redundant robotic manipulators based on visual data [16,17]. In [18], a medium size neural network is combined with MPC and reinforcement learning to produce stable gaits and locomotion tasks.

NMPC [19] and learning-based control, such as reinforcement learning [20], are methods for optimal control of dynamic systems which have evolved in parallel in the control systems and computational intelligence

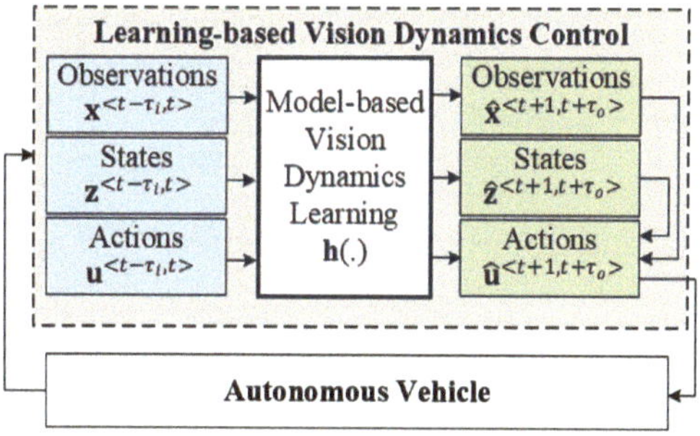

Figure 10.1 ***Vision dynamics-based learning control.*** Given sequences of past observations $\mathbf{x}^{<t-\tau_i,t>}$, states $\mathbf{z}^{<t-\tau_i,t>}$ and control inputs $\mathbf{u}^{<t-\tau_i,t>}$, the goal is to learn a visual dynamics model $\mathbf{h}(\cdot)$ that can be used to compute an optimal control strategy along prediction horizon $[t+1, t+\tau_O]$.

communities, respectively. Throughout the paper, we use a notation which brings together both communities. The value of a variable is defined either for a single discrete time step t, written as superscript $<t>$, or as a discrete sequence defined in the $<t, t+k>$ time interval, where k denotes the length of the sequence. For example, the value of a state variable $\mathbf{z}$ is defined either at discrete time t, as $\mathbf{z}^{<t>}$, either within a sequence interval $\mathbf{z}^{<t,t+k>}$. Vectors and matrices are indicated by bold symbols.

10.2 Problem definition

The block diagram of a vision dynamics based control system is shown in Fig. 10.1. Given sequences of past observations $\mathbf{x}^{<t-\tau_i,t>}$, vehicle states $\mathbf{z}^{<t-\tau_i,t>}$ and control inputs $\mathbf{u}^{<t-\tau_i,t>}$, the task is to learn a vision dynamics model $\mathbf{h}(\cdot)$, such that predictions on future observations $\mathbf{x}^{<t+1,t+\tau_o>}$, states $\mathbf{z}^{<t+1,t+\tau_o>}$ and control actions $\mathbf{u}^{<t+1,t+\tau_o>}$ are as close as possible to their true distribution. Given a discrete sampling time t, we define τ_i and τ_o as past and future temporal horizons, respectively.

Consider the following nonlinear, state-space system:

$$\mathbf{z}^{<t+1>} = \mathbf{f}_{true}(\mathbf{z}^{<t>}, \mathbf{u}^{<t>}), \tag{10.1}$$

with observable state $\mathbf{z}^{<t>} \sim \mathcal{N}(\mathbf{z}^{<t>}, \Sigma^{<t>})$, $\mathbf{z}^{<t>} \in \mathbb{R}^n$ and control input $\mathbf{u}^{<t>} \in \mathbb{R}^m$, at discrete time t. The true system $\mathbf{f}_{true}$ is not known exactly and is approximated by the sum of an *a priori* model and a learned vision dynamics model:

$$\mathbf{z}^{<t+1>} = \underbrace{\mathbf{f}(\mathbf{z}^{<t>}, \mathbf{u}^{<t>})}_{a\ priori\ \text{model}} + \underbrace{\mathbf{h}(\mathbf{s}^{<t-\tau_i,t>})}_{\text{vision dynamics model}}, \tag{10.2}$$

with environmental and disturbance dependencies $\mathbf{s}^{<t>} \in \mathbb{R}^p$:

$$\mathbf{s}^{<t>} = (\mathbf{z}^{<t-\tau_i,t>}, \mathbf{x}^{<t-\tau_i,t>}) \tag{10.3}$$

$\mathbf{s}^{<t>}$ is defined as the system's state and the measured state of the environment, both at time t. $\mathbf{s}^{<t>}$ represents the set of historic dependencies integrated along time interval $[t - \tau_i, t]$.

The models $\mathbf{f}(\cdot)$ and $\mathbf{h}(\cdot)$ are nonlinear process models. $\mathbf{f}(\cdot)$ is a known process model, representing our knowledge of $\mathbf{f}_{true}(\cdot)$, and $\mathbf{h}(\cdot)$ is a learned vision dynamics model, representing discrepancies between the response of the *a priori* model and the optimal behavior of the system. The behavior and the initially unknown disturbance models are modeled (Section 10.2.1) as a deep neural network which estimates the optimal behavior of the system in different cases which cannot be modeled *a priori*.

The role of the vision dynamics model is to estimate the desired future states of the system. In this sense, we distinguish between a given vehicle route $\mathbf{z}_{ref}^{<t-\infty,t+\infty>}$, which, from a control perspective, is practically infinite, and a desired policy $\mathbf{z}_d^{<t+1,t+\tau_o>}$ which is estimated over a finite time horizon τ_o. $\mathbf{z}_d^{<t+1,t+\tau_o>}$ is a quantitative deviation of the system from the reference trajectory, required in order to cope with an unpredictable event. Thus, by requiring as reference only a global rough state trajectory, we manage to estimate in a closed-loop manner the desired response of the NMPC controller. For the autonomous driving case, this translates to the route that the vehicle should follow from its start location to destination. We propose to use a deep recurrent neural network to estimate $\mathbf{h}(\cdot)$ and to calculate the desired policy $\mathbf{z}_d^{<t+1,t+\tau_o>}$ over prediction horizon $[t+1, t+\tau_o]$.

On top of the vision dynamics model, we define the cost function to be optimized by the NMPC in discrete time interval $[t+1, t+\tau_o]$ as

$$J(\mathbf{z}, \mathbf{u}) = (\mathbf{z}_d - \mathbf{z})^T \mathbf{Q} (\mathbf{z}_d - \mathbf{z}) + \mathbf{u}^T \mathbf{R} \mathbf{u}, \tag{10.4}$$

where $\mathbf{Q} \in \mathbb{R}^{\tau_o n \times \tau_o n}$ is positive semi-definite, $\mathbf{R} \in \mathbb{R}^{\tau_o M \times \tau_o M}$ is positive definite, $\mathbf{z}_d^{<t+1,t+\tau_o>} = [\mathbf{z}_d^{<t+1>}, ..., \mathbf{z}_d^{<t+\tau_o>}]$ is a sequence of desired states estimated by the model, $\mathbf{z}^{<t+1,t+\tau_o>} = [\mathbf{z}^{<t+1>}, ..., \mathbf{z}^{<t+\tau_o>}]$ is the sequence of predicted states and $\mathbf{u}^{<t,t+\tau_o-1>} = [\mathbf{u}^{<t>}, ..., \mathbf{u}^{<t+\tau_o-1>}]$ is the control input sequence.

The constrained NMPC objective is to find a set of control actions which optimize the plant's behavior over a give time horizon τ_o, while satisfying a set of hard and soft constraints:

$$(\mathbf{z}_{opt}^{<t+1>}, \mathbf{u}_{opt}^{<t+1>}) = \arg\min_{\mathbf{z},\mathbf{u}} J\ (\mathbf{z}^{<t+1,t+\tau_o>}, \mathbf{u}^{<t+1,t+\tau_o>}) \tag{10.5a}$$

$$\text{such that } \mathbf{z}^{<0>} = \mathbf{z}^{<t>} \tag{10.5b}$$

$$\begin{aligned}\mathbf{z}^{<t+i+1>} =& \mathbf{f}(\mathbf{z}^{<t>}, \mathbf{u}^{<t>}) + \\ & \mathbf{h}(\mathbf{s}^{<t-\tau_i,t>}) + \\ & \mathbf{g}(\mathbf{s}^{<t-\tau_i,t>}),\end{aligned} \tag{10.5c}$$

$$\mathbf{e}_{\min}^{<t+i>} \leq \mathbf{e}^{<t+i>} \leq \mathbf{e}_{\max}^{<t+i>}, \tag{10.5d}$$

$$\mathbf{u}_{\min}^{<t+i>} \leq \mathbf{u}^{<t+i>} \leq \mathbf{u}_{\max}^{<t+i>}, \tag{10.5e}$$

$$\dot{\mathbf{u}}_{\min}^{<t+i>} \leq \frac{\dot{\mathbf{u}}^{<t+i>} - \dot{\mathbf{u}}^{<t+i-1>}}{\Delta t} \leq \dot{\mathbf{u}}_{\max}^{<t+i>}, \tag{10.5f}$$

where $i = 0, 1, ..., \tau_o - 1$, $\mathbf{z}^{<0>}$ is the initial state and Δt is the sampling time of the controller. $\mathbf{e}^{<t+i>} = \mathbf{z}_d^{t+i} - \mathbf{z}^{t+i}$ is the cross-track error, $\mathbf{e}_{min}^{<t+i>}$ and $\mathbf{e}_{max}^{<t+i>}$ are the lower and upper tracking bounds, respectively. Additionally, we consider $\mathbf{u}_{\min}^{<t+i>}$, $\dot{\mathbf{u}}_{\min}^{<t+i>}$ and $\mathbf{u}_{\max}^{<t+i>}$, $\dot{\mathbf{u}}_{\max}^{<t+i>}$ as lower and upper constraint bounds for the actuator and actuator rate of change, respectively. The NMPC controller implements

$$\mathbf{u}^{<t>} = \mathbf{u}_{opt}^{<t+1>}, \tag{10.6}$$

at each iteration t.

We leverage on the quadratic cost function (10.4) and solve the nonlinear optimization problem described above using the Broyden–Fletcher–Goldfarb–Shanno algorithm [21]. The quadratic form allows us to apply the quasi-Newton optimization method, without the need to specify the Hessian matrix.

10.2.1 Learning a vision dynamics model

The role of the vision dynamics model is to estimate the optimal desired state trajectory $\mathbf{z}_d^{<t+1,t+\tau_o>}$ using temporal information and the given reference trajectory $\mathbf{z}_{ref}^{<t-\infty,t+\infty>}$. For practical reasons, we consider the reference trajectory within the finite time interval $[t-\tau_i, t+\tau_o]$. In the following, we formulate the vision dynamics modeling problem in an inverse reinforcement learning setup for an autonomous agent.

Given a sequence of temporal observations $\mathbf{x}^{<t-\tau_i,t>} : \mathbb{R}^w \times \tau_i \rightarrow \mathbb{R}^w \times \tau_o$, the system's state $\mathbf{z}^{<t>} \in \mathbb{R}^n$ in $\mathbf{I}^{<t>}$ and the reference set-points $\mathbf{z}_{ref}^{<t+\tau_o>} \in \mathbb{R}^n$ in observation space at time t, the task is to learn a compensating behavior policy for navigating from state $\mathbf{z}^{<t>}$ to the destination state $\mathbf{z}_{ref}^{<t+\tau_o>}$, where τ_o is the number of time steps for which the set-point trajectory is planned.

In reinforcement learning terminology, the above problem can be modeled as a *Markov Decision Process* (MDP) $M := (S, Z_d, \mathcal{T}, R, \gamma)$, where:

- S represents a finite set of states, $\mathbf{s}^{<t>} \in S$ being the state of the agent at time t. The state is defined as the pair $\mathbf{s}^{<t>} = (\mathbf{z}^{<t-\tau_i,t>}, \mathbf{x}^{<t-\tau_i,t>})$.
- Z_d represents a finite set of set-point sequences allowing the agent to navigate through the environment defined by $\mathbf{x}^{<t-\tau_i,t>}$, where $\mathbf{z}_d^{<t+1,t+\tau_o>} \in Z_d$ is the future predicted optimal behavior policy that the agent should perform in the next time interval $[t+1, t+\tau_o]$. A behavioral policy $\mathbf{z}_d^{<t+1,t+\tau_o>}$ is defined as the collection of estimated trajectory set-points which are used by the NMPC to compute the optimal control actions.
- $\mathcal{T} : S \times Z_d \times S \rightarrow [0, 1]$ is a stochastic transition function, where $\mathcal{T}_{\mathbf{s}^{<t>}, \mathbf{z}_d^{<t+1,t+\tau_o>}}^{\mathbf{s}^{<t+\tau_o>}}$ describes the probability of arriving in state $\mathbf{s}^{<t+\tau_o>}$, after performing the behavioral actions $\mathbf{z}_d^{<t+1,t+\tau_o>}$ in state $\mathbf{s}^{<t>}$.
- $R : S \times A \times S \rightarrow \mathbb{R}$ is a scalar reward function which controls the estimation of $\mathbf{z}_d^{<t+1,t+\tau_o>}$, where $R_{\mathbf{s}^{<t>}, \mathbf{z}_d^{<t+1,t+\tau_o>}}^{\mathbf{s}^{<t+\tau_o>}} \in \mathbb{R}$. For a state transition $\mathbf{s}^{<t>} \rightarrow \mathbf{s}^{<t+\tau_o>}$ at time t, we define

$$R_{\mathbf{s}^{<t>}, \mathbf{z}_d^{<t+1,t+\tau_o>}}^{\mathbf{s}^{<t+\tau_o>}} = ||\mathbf{z}_d^{<t>} - \mathbf{z}_{ref}^{<t+\tau_o>}||_2 - \\ - ||\mathbf{z}_d^{<t+1>} - \mathbf{z}_{ref}^{<t+\tau_o>}||_2, \tag{10.7}$$

  where $||\cdot||_2$ is the L2 norm. The reward function is a distance feedback, which is smaller if the desired system's state follows a minimal energy trajectory to the reference state $\mathbf{z}_{ref}^{<t+\tau_o>}$ and large otherwise.
- γ is the discount factor controlling the importance of future versus immediate rewards.

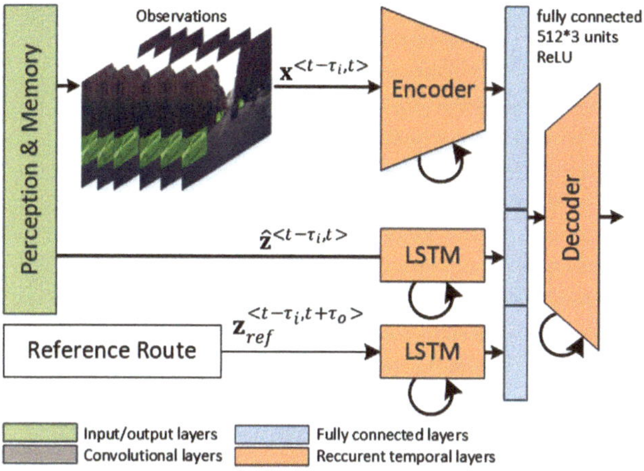

Figure 10.2 ***Deep Q-network architecture for vision dynamics learning.*** The training data consists of observation sequences $\mathbf{x}^{<t-\tau_i,t>}$, historic system states $\mathbf{z}^{<t-\tau_i,t>}$, and reference state trajectories $\mathbf{z}_{ref}^{<t-\tau_i,t+\tau_o>}$, together with their respective desired states $\mathbf{z}_d^{<t+1,t+\tau_o>}$. Before being stacked upon each other, the observations stream is passed through a convolutional encoding neural network. The merged data is further fed to an LSTM decoder via three fully connected layers of 512 units, respectively. During runtime, the optimal action-value function $Q^*(\mathbf{s}, \mathbf{z}_d)$ and the desired state trajectory $\mathbf{z}_d^{<t+1,t+\tau_o>}$ is computed solely from observation sequences and historic state information.

Considering the proposed reward function and an arbitrary set-point trajectory $T = [\mathbf{z}_d^{<0>}, \mathbf{z}_d^{<1>}, ..., \mathbf{z}_d^{<k>}]$ in observation space, at any time $\hat{t} \in [0, 1, ..., k]$, the associated cumulative future discounted reward is defined as

$$R^{<\hat{t}>} = \sum_{t=\hat{t}}^{k} \gamma^{<t-\hat{t}>} r^{<t>}, \tag{10.8}$$

where the immediate reward at time t is given by $r^{<t>}$. In reinforcement learning theory, the statement in Eq. (10.8) is known as a finite horizon learning episode of sequence length k [20].

The behavioral model's objective is to find the desired set-point policy that maximizes the associated cumulative future reward. We define the optimal action-value function $Q^*(\cdot, \cdot)$ which estimates the maximal future discounted reward when starting in state $\mathbf{s}^{<t>}$ and performing the NMPC control actions $\mathbf{u}^{<t+1,t+\tau_o>}$, given an estimated policy set-point $\mathbf{z}_d^{<t+1,t+\tau_o>}$:

$$Q^*(\mathbf{s}, \mathbf{z}_d) = \max_{\pi} \mathbb{E}\ [R^{<\hat{t}>} | \mathbf{s}^{<\hat{t}>} = \mathbf{s},\ \mathbf{z}_d^{<t+1,t+\tau_o>} = \mathbf{z}_d,\ \pi], \tag{10.9}$$

where π is a behavioral policy, or action, which is a probability density function over a set of possible actions that can take place in a given state. The optimal action-value function $Q^*(\cdot, \cdot)$ maps a given state to the optimal behavior policy of the agent in any state:

$$\forall \mathbf{s} \in S : \pi^*(\mathbf{s}) = \arg\max_{\mathbf{z}_d \in Z_d} Q^*(\mathbf{s}, \mathbf{z}_d). \tag{10.10}$$

The optimal action-value function Q^* satisfies the Bellman optimality equation [22], which is a recursive formulation of Eq. (10.9):

$$\begin{aligned} Q^*(\mathbf{s}, \mathbf{z}_d) &= \sum_{\mathbf{s}} \mathcal{T}_{\mathbf{s},\mathbf{z}_d}^{\mathbf{s}'} \left(R_{\mathbf{s},\mathbf{z}_d}^{\mathbf{s}'} + \gamma \cdot \max_{\mathbf{z}'_d} Q^*(\mathbf{s}', \mathbf{z}'_d) \right) \\ &= \mathbb{E}_{\mathbf{z}'_d} \left(R_{\mathbf{s},\mathbf{z}_d}^{\mathbf{s}'} + \gamma \cdot \max_{\mathbf{z}'_d} Q^*(\mathbf{s}', \mathbf{z}'_d) \right), \end{aligned} \tag{10.11}$$

where $\mathbf{z}_d = \mathbf{z}_d^{<t+1,t+\tau_o>}$, $\mathbf{s}' = \mathbf{s}^{<t+\tau_o>}$ represents a possible state visited after $\mathbf{s} = \mathbf{s}^{<t>}$ and $\mathbf{z}'_d = \mathbf{z}_d^{\prime<t+1,t+\tau_o>}$ is the corresponding behavioral policy, or action. The model-based policy iteration algorithm was introduced in [20], based on the proof that the Bellman equation is a contraction mapping [23] when written as an operator ν:

$$\forall Q, \lim_{n \to \infty} \nu^{(n)}(Q) = Q^*. \tag{10.12}$$

However, the standard reinforcement learning method described above is not feasible due to the high dimensional observations space. In autonomous driving applications, the observation space is mainly composed of sequences of sensory information made up of images, radar, LiDAR, etc. Instead of the traditional approach, we use a non-linear parametrization of Q^* for autonomous driving, encoded in the deep neural network illustrated in Fig. 10.2. In the literature, such a non-linear approximator is called a Deep Q-Network (DQN) [6] and is used for estimating the approximate action-value function:

$$Q(\mathbf{s}^{<t>}, \mathbf{z}_d^{<t+1,t+\tau_o>}; \Theta) \approx Q^*(\mathbf{s}^{<t>}, \mathbf{z}_d^{<t+1,t+\tau_o>}), \tag{10.13}$$

where $\Theta = [\mathbf{W}_i, \mathbf{U}_i, \mathbf{b}_i]$ are the parameters of the Deep Q-Network.

In the deep Q-network from Fig. 10.2, the environment observations are firstly passed through a series of convolutional, ReLU activations and max pooling layers. This builds an abstract representation which is stacked on top of the previous system's states $\mathbf{z}^{<t-\tau_i,t>}$ and reference trajectory $\mathbf{z}_{ref}^{<t-\tau_i,t+\tau_o>}$. The stacked representation is processed by a fully connected neural network layer, before being fed as input to an LSTM network. The final output returns the optimal action-value function $Q^*(\cdot,\cdot)$, together with the desired policy $\mathbf{z}_d = \mathbf{z}_d^{<t+1,t+\tau_o>}$, which represents the desired input to the constrained NMPC controller.

By taking into account the Bellman optimality equation (10.11), it is possible to train a deep Q-network in an inverse reinforcement learning manner through the minimization of the mean squared error. The optimal expected Q value can be estimated within a training iteration i based on a set of reference parameters $\bar{\Theta}_i$ calculated in a previous iteration i':

$$\mathbf{z}_d = R_{\mathbf{s},\mathbf{z}_d}^{\mathbf{s}'} + \gamma \cdot \max_{\mathbf{z}'_d} Q(\mathbf{s}', \mathbf{z}'_d; \bar{\Theta}_i), \tag{10.14}$$

where $\bar{\Theta}_i := \Theta_{i'}$. The new estimated network parameters at training step i are evaluated using the following squared error function:

$$\hat{\Theta}_i = \min_{\Theta_i} \mathbb{E}_{\mathbf{s},\mathbf{z}_d,r,\mathbf{s}'}\left[(\mathbf{z}_d - Q(\mathbf{s}, \mathbf{z}_d; \Theta_i))^2\right], \tag{10.15}$$

where $r = R_{\mathbf{s},\mathbf{z}_d}^{\mathbf{s}'}$. Based on (10.15), we apply maximum likelihood estimation for calculating the weights of the deep Q-network. The gradient is approximated with random samples in the backpropagation through time algorithm, which uses stochastic gradient descent for training:

$$\nabla_{\Theta_i} = \mathbb{E}_{\mathbf{s},\mathbf{z}_d,r,\mathbf{s}'}\left[(\mathbf{z}_d - Q(\mathbf{s}, \mathbf{z}_d; \Theta_i))\, \nabla_{\Theta_i}(Q(\mathbf{s}, \mathbf{z}_d; \Theta_i))\right]. \tag{10.16}$$

In comparison to traditional DRL setups, where the actions space consisted on only a couple of actions, such as turn left, turn right, accelerate, decelerate, the action space in our approach is much larger and depends on the prediction horizon τ_o.

10.3 Experiments

A basic illustration of the autonomous driving problem space is shown in Fig. 10.3. Given a reference trajectory route $\mathbf{z}^{<t-\infty,t+\infty>}$, a sequence of observations $\mathbf{x}^{<t-\tau_i,t>}$ and past vehicle states $\mathbf{z}^{<t-\tau_i,t>}$, we want to estimate the optimal driving control action $\mathbf{u}_{opt}^{<t+1>}$ at time $t+1$, such that the vehicle follows an optimal future states trajectory $\mathbf{z}^{<t+1,t+\tau_o>}$. The optimal future states are calculated by the constrained NMPC controller, based on the desired states $\mathbf{z}_d^{<t+1,t+\tau_o>}$ estimated by the vision dynamics model $\mathbf{h}(\cdot)$.

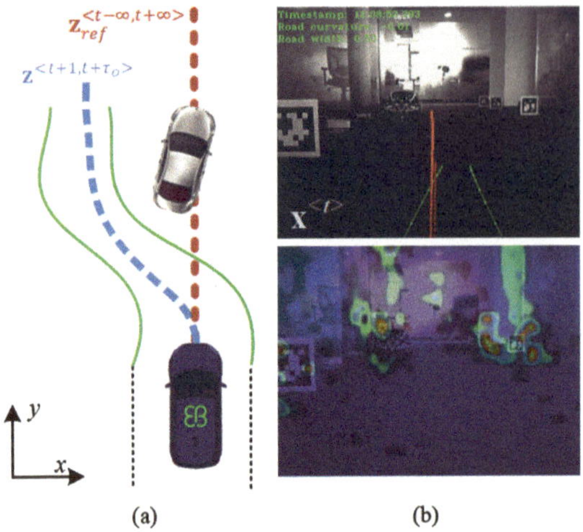

Figure 10.3 ***Autonomous driving problem space (a) and the visual representation of the vision dynamics model (b).*** Given the current state of the vehicle $\mathbf{z}^{<t>}$, a reference trajectory $\mathbf{z}_{ref}^{<t-\infty,t+\infty>}$ and an input sequence of observation $\mathbf{x}^{<t-\tau_i,t>}$, the goal is to estimate the optimal future vehicle states trajectory $\mathbf{z}^{<t+1,t+\tau_i>}$, over control horizon τ_o.

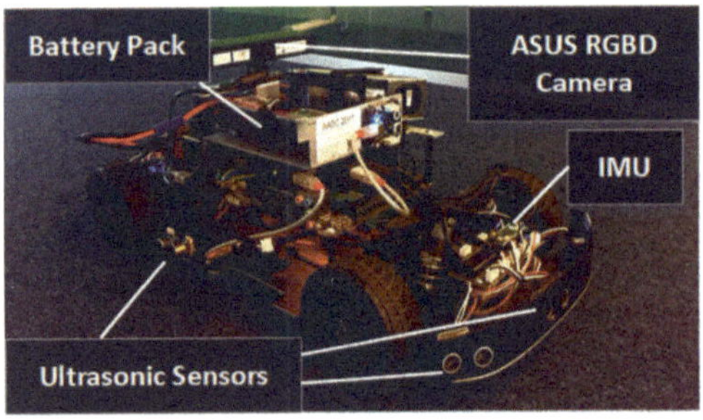

Figure 10.4 ***Model-car test vehicles used in the experiments.***

The experiments compared three algorithms: Dynamic Window Approach (DWA) [24], End2End learning control and the Vision Dynamics (VD) approach. The real world Audi model-car vehicle [25] from Fig. 10.4 was used for performance evaluation, with the task to safely navigate a 120 m loop in an indoor environment, which contains six intersections.

The test vehicle is 60 cm long, 30 cm wide, and 25 cm high, with differential carriage. The car is driven by a Hacker Skalar 10 brushless motor and an underling VESC motor controller. An Asus Xtion Pro Live front-facing RGB-Depth camera and 9 ultrasonic sensors are used to sense the

Table 10.1 Model-car parameters.

Description	Value
Dimensions	60 cm × 30 cm × 25 cm
Chassis	Differential carriage
Drive	Hacker Skalar 10 brushless motor
	1x Asus Xtion Pro Live RGBD camera
Sensing	9 ultrasonic sensors
	1x right side LiDAR
Main computing device	2x Arduino MICROS6
Odometry	IMU Brick v2
Operating System	Ubuntu Linux 14.04.3
Deployment SW	Elektrobit ADTF v2.13.2 x64

environment. 5 ultrasonic sensors are located in the front bumper, three in the rear bumper and one sensor at the left side of the car. The right side features a non-rotating single beam LiDAR. All sensors are connected to two Arduino MICROS6 boards. The microcontrollers trigger measurements, perform preprocessing and deliver the data via USB. A further microcontroller measures the voltage level of the batteries. An additional Arduino board is used to communicate with the motor controller. The odometry is based on an IMU Brick v2 sensor. The model-car test setup parameters are shown in Table 10.1.

The model car's operating system runs with the Elektrobit Automotive Data and Time-Triggered Framework v2.13.2 (ADTF), on top of an Ubuntu 14.04.3 Linux distribution. For processing the data stream and calculating the car's control commands, we use the desktop training computer, connected to the car via a wireless communication network. For gathering training data, we ran the model car on the 3rd floor of the Elektrobit building for 2 hours, gathering 79.200 training observations with the corresponding vehicle states.

Once trained, the algorithms have to solve the goal navigation task in the same driving environment for 10 different trials. The mean of the driving trajectories acquired during training data acquisition is considered as the ground truth path. Five trials contained no obstacles on the ground truth path, while the other five had static and dynamic obstacles (humans) on their reference trajectory. We show the maximum tracking errors and travel times vs. trial number in Fig. 10.5. The End2End results have the largest heading tracking error. One reason for this phenomenon is the discrete steering commands (e.g. turn left, turn right, accelerate, decelerate)

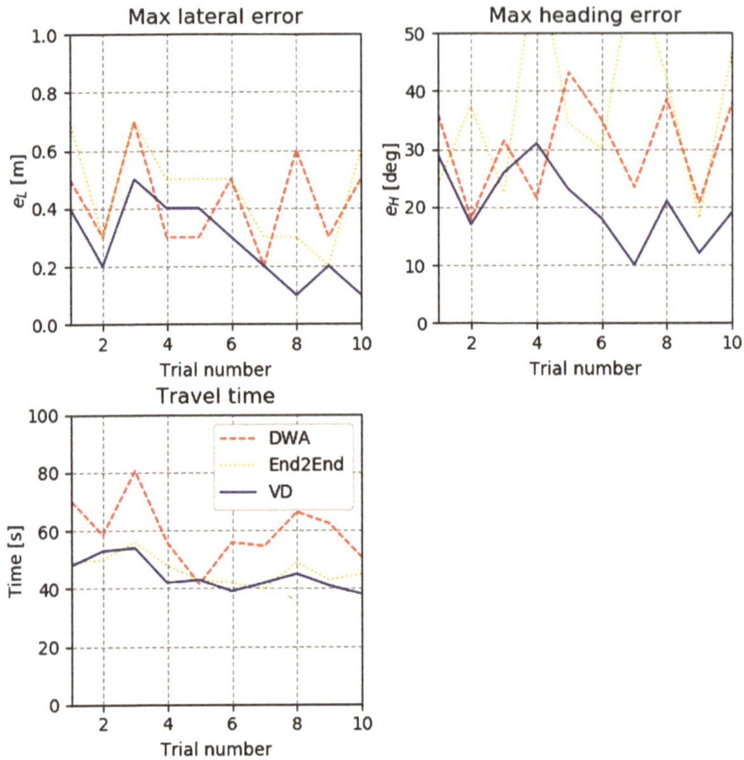

Figure 10.5 ***Maximum path tracking errors and travel times vs. trial number.*** VD results in lower tracking errors and an increased travel time for the 10 experimental trials.

provided as control output by the End2End method. In comparison, VD computes a smoother trajectory, yielding lower path tracking errors.

Fig. 10.6 shows path tracking errors and control inputs for a sample distance of 10 m. The effect of the discrete steering commands calculated by the End2End controller is visible in the tracking errors as the jittering phenomenon. This confirmed our objective to design a learning based vision dynamics NMPC that can predict and adapt the trajectory of an ego-vehicle smoothly, according to the configuration of the driving environment.

10.4 Conclusions

In this work, we have introduced a novel vision dynamics learning control method for autonomous vehicles operating in different driving conditions. The controller makes use of an a priori process model and a vision dy-

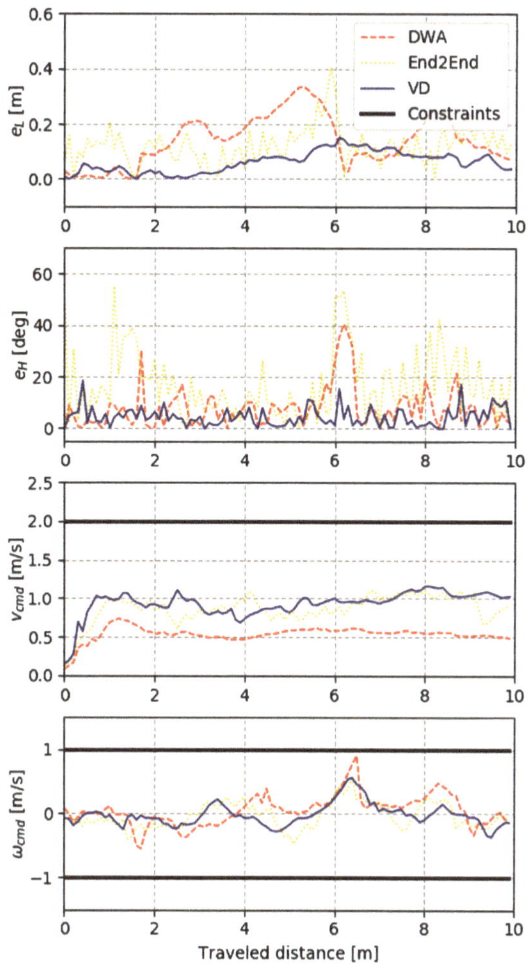

Figure 10.6 ***Path tracking errors and steering commands vs. a 10 m travel distance.*** The VD algorithm outputs a smoother trajectory, resulting in lower path tracking errors.

namics model encoded in the layers of a deep neural network. We show that by incorporating the dynamics of the scene in a deep network, the vision dynamics approach can be used to steer an autonomous vehicle in different operating conditions, such as cluttered indoor environments. We train our system in an inverse reinforcement learning fashion, based on the Bellman optimality principle. As future work, the authors plan to investigate the stability of the vision dynamics controller, especially in relation to the functional safety requirements needed for deployment on a larger scale.

Through high robustness, the proposed method might represent an important component of next-generation autonomous driving technologies that contribute to safer and better self-driving cars.

References

[1] X. Geng, H. Liang, B. Yu, P. Zhao, L. He, R. Huang, A scenario-adaptive driving behavior prediction approach to urban autonomous driving, Applied Sciences 7 (4) (Apr 2017).

[2] M. Bojarski, D.D. Testa, D. Dworakowski, B. Firner, B. Flepp, P. Goyal, L.D. Jackel, M. Monfort, U. Muller, J. Zhang, X. Zhang, J. Zhao, K. Zieba, End to end learning for self-driving cars, CoRR, arXiv:1604.07316.

[3] Y. Pan, C.-A. Cheng, K. Saigol, K. Lee, X. Yan, E. Theodorou, B. Boots, Imitation learning for agile autonomous driving, International Journal of Robotics Research (2019).

[4] B. Okal, K.O. Arras, Learning socially normative robot navigation behaviors with Bayesian inverse reinforcement learning, in: Int. Conf. on Robotics and Automation ICRA 2016, IEEE, 2016, pp. 2889–2895.

[5] E.D. Dickmanns, B.D. Mysliwetz, Recursive 3-D road and relative ego-state recognition, IEEE Transactions on Pattern Analysis and Machine Intelligence 14 (2) (1992) 199–213.

[6] V. Mnih, K. Kavukcuoglu, D. Silver, A.A. Rusu, J. Veness, M.G. Bellemare, A. Graves, M. Riedmiller, A.K. Fidjeland, G. Ostrovski, S. Petersen, C. Beattie, A. Sadik, I. Antonoglou, H. King, D. Kumaran, D. Wierstra, S. Legg, D. Hassabis, Human-level control through deep reinforcement learning, Nature 518 (7540) (2015) 529–533.

[7] G. Williams, P. Drews, B. Goldfain, J.M. Rehg, E.A. Theodorou, Aggressive driving with model predictive path integral control, in: Int. Conf. on Robotics and Automation (ICRA), 2016, pp. 1433–1440.

[8] C.E. Rasmussen, Gaussian Processes for Machine Learning, MIT Press, 2006.

[9] D. Nguyen-Tuong, M. Seeger, J. Peters, Local Gaussian process regression for real time online model learning, in: Proceedings of the Neural Information Processing Systems Conference, 2008, pp. 1193–1200.

[10] F. Meier, P. Hennig, S. Schaal, Efficient Bayesian local model learning for control, in: IEEE/RSJ International Conference on Intelligent Robots and Systems (IROS) 2016, IEEE, 2014, pp. 2244–2249.

[11] C.J. Ostafew, J. Collier, A.P. Schoellig, T.D. Barfoot, Learning-based nonlinear model predictive control to improve vision-based mobile robot path tracking, Journal of Field Robotics 33 (1) (2015) 133–152.

[12] C.J. Ostafew, A.P. Schoellig, T.D. Barfoot, Robust constrained learning-based NMPC enabling reliable mobile robot path tracking, International Journal of Robotics Research 35 (13) (2016) 1547–1563.

[13] O. Sigaud, C. Salaün, V. Padois, On-line regression algorithms for learning mechanical models of robots: a survey, Robotics and Autonomous Systems 59 (12) (2011) 1115–1129.

[14] K.J. Hunt, D. Sbarbaro, R. Zbikowski, P.J. Gawthrop, Neural networks for control systems—a survey, Automatica (1992).

[15] G. Bekey, K.Y. Goldberg, Neural Networks in Robotics, Springer US, 1992.

[16] N. Wahlstrom, T.B. Schon, M.P. Deisenroth, From pixels to torques: policy learning with deep dynamical models, in: Deep Learning Workshop at ICML, 2015.

[17] N. Mishra, P. Abbeel, I. Mordatch, Prediction and control with temporal segment models, in: ICML, 2017.

[18] A. Nagabandi, G. Kahn, R.S. Fearing, S. Levine, Neural network dynamics for model-based deep reinforcement learning with model-free fine-tuning, in: Int. Conf. on Robotics and Automation ICRA 2018, Brisbane, Australia, 2018.
[19] D.Q. Mayne, J.B. Rawlings, C.V. Rao, P.O.M. Scokaert, Constrained model predictive control: stability and optimality, Automatica 36 (2000) 789–814.
[20] R. Sutton, A. Barto, Introduction to Reinforcement Learning, MIT Press, 1998.
[21] R. Fletcher, Practical Methods of Optimization, 2nd edition, John Wiley & Sons, New York, NY, USA, 1987.
[22] R. Bellman, Dynamic Programming, Princeton University Press, 1957.
[23] C. Watkins, P. Dayan, Q-learning, Machine Learning 8 (3) (1992) 279–292.
[24] D. Fox, W. Burgard, S. Thrun, The dynamic window approach to collision avoidance, IEEE Robotics & Automation Magazine 4 (1) (1997) 23–33.
[25] Audi autonomous driving cup, https://www.audi-autonomous-driving-cup.com/, 2018.

Index

R

S

W

Z